Research Report

T0302508

Assessment of Beddown Alternatives for the F-35

Ronald G. McGarvey, James H. Bigelow, Gary James Briggs, Peter Buryk, Raymond E. Conley, John G. Drew, Perry Shameem Firoz, Julie Kim, Lance Menthe, S. Craig Moore, William W. Taylor, William A. Williams

RAND Project AIR FORCE

Prepared for the United States Air Force
Approved for public release; distribution unlimited

RAND CORPORATION

The research described in this report was sponsored by the United States Air Force under Contract FA7014-06-C-0001. Further information may be obtained from the Strategic Planning Division, Directorate of Plans, Hq USAF.

Library of Congress Cataloging-in-Publication Data

McGarvey, Ronald G.
Assessment of beddown alternatives for the F-35 / Ronald G. McGarvey, James H. Bigelow, Gary James Briggs, Peter Buryk, Raymond E. Conley, John G. Drew, Perry Shameem Firoz, Julie Kim, Lance Menthe, S. Craig Moore, William W. Taylor, William A. Williams.
pages cm
Includes bibliographical references.
ISBN 978-0-8330-7807-0 (pbk. : alk. paper)
1. F-35 (Jet fighter plane)—Cost control. 2. United States. Air Force—Reorganization. 3. United States. Air Force—Appropriations and expenditures. I. Title.

UG1242.F5M397 2013
358.4'383—dc23

2013005932

Published 2013 by the RAND Corporation
1776 Main Street, P.O. Box 2138, Santa Monica, CA 90407-2138
1200 South Hayes Street, Arlington, VA 22202-5050
4570 Fifth Avenue, Suite 600, Pittsburgh, PA 15213-2665
RAND URL: http://www.rand.org/
To order RAND documents or to obtain additional information, contact
Distribution Services: Telephone: (310) 451-7002;
Fax: (310) 451-6915; Email: order@rand.org

Preface

As currently planned, the F-35 Joint Strike Fighter is the largest aircraft acquisition program in the history of the Department of Defense (DoD). According to the December 2011 F-35 Selected Acquisition Report (DoD, 2011), the total acquisition cost to procure 2,457 F-35 aircraft across the United States Air Force (USAF), Navy, and Marine Corps is $331 billion, with total operating and support (O&S) costs of $617 billion to operate the aircraft through 2065 (both costs are computed using a base year of 2012). Moreover, the F-35 cost-per-flying-hour estimate has increased by more than 80 percent (in constant dollars) over the interval 2002 to 2010. To ensure that the affordability of the F-35 program is not threatened by continuing O&S cost growth, the USAF is examining alternative strategies to reduce those costs.

One approach to reducing O&S costs is to reduce the number of F-35 home-station operating locations, which is in turn related to the number of F-35 squadrons and the number of Primary Aerospace Vehicle Authorized (PAA) per squadron. In December 2011, RAND Project AIR FORCE (PAF) presented a preliminary analysis to the Director of Logistics, Office of the Deputy Chief of Staff for Logistics, Installations and Mission Support, Headquarters USAF (AF/A4L); this analysis found that increasing the combat-coded PAA per squadron while maintaining a constant total number of PAA (thereby reducing the total number of squadrons) could significantly reduce maintenance manpower and support equipment costs.

Based upon these findings, the Vice Chief of Staff of the Air Force asked PAF to assess whether savings could be achieved by reconfiguring the USAF's 960 combat-coded F-35 PAA into larger squadrons (i.e., increasing PAA per squadron), by adjusting the PAA mix across the Active Component and Reserve Component, and by adjusting the percentage of Active Component PAA assigned to continental U.S. home-station locations. This report addresses how such changes would affect the USAF in the following ways:

- Ability to support both surge and steady-state contingency operations
- Ability to absorb the necessary number of F-35 pilots
- Requirements for maintenance manpower and support equipment
- Requirements for new infrastructure across the set of existing F-16 and A-10 bases
- Ability to develop future senior leaders out of the pool of fighter pilots.

A companion executive summary volume is also available:

> Ronald G. McGarvey et al., *Assessment of Beddown Alternatives for the F-35: Executive Summary*, RR-124/1-AF, 2013.

This research was conducted within the Resource Management Program of RAND PAF for two fiscal year 2012 projects, "Reducing F-35 Operations and Sustainment Costs" and "Identifying Potential Efficiencies in the F-35 Basing Posture," sponsored at that time by, respectively, Major General Judith Fedder, HQ AF/A4L, and General Philip Breedlove, the Vice Chief of Staff of the Air Force.

This report should be of interest to operations planners, logisticians, and manpower personnel throughout the USAF.

The views expressed are those of the authors and do not necessarily reflect the official policy or position of the Department of the Air Force or the U.S. government.

RAND Project AIR FORCE

RAND Project AIR FORCE (PAF), a division of the RAND Corporation, is the U.S. Air Force's federally funded research and development center for studies and analyses. PAF provides the Air Force with independent analyses of policy alternatives affecting the development, employment, combat readiness, and support of current and future air, space, and cyber forces. Research is conducted in four programs: Force Modernization and Employment; Manpower, Personnel, and Training; Resource Management; and Strategy and Doctrine.

Additional information about PAF is available on our website:
http://www.rand.org/paf/

Contents

Figures

Tables

Summary

As currently planned, the F-35 Joint Strike Fighter is the largest aircraft acquisition program in Department of Defense history. To ensure that the affordability of the F-35 program is not threatened by continuing operating and support (O&S) cost growth, the U.S. Air Force (USAF) is examining alternative strategies to reduce those costs.

One approach to reducing O&S costs is to increase the number of Primary Aerospace Vehicle Authorized (PAA) per squadron, across a constant total number of USAF PAA, with a resulting reduction in the number of F-35 home-station operating locations. In 2012, the commander of Air Combat Command (ACC/CC) approved a beddown plan to determine how to allocate the 960 combat-coded PAA across fighter squadrons and operating locations. The plan calls for the aircraft to be allocated into squadrons of 24 PAA in the Active Component (AC) and Air Force Reserve Command (AFRC), and 18 PAA per squadron in the Air National Guard (ANG). A total of 44 squadrons would be distributed among 31 operating locations.

At the request of the Vice Chief of Staff of the Air Force, RAND Project Air Force (PAF) assessed whether O&S savings could be achieved by (1) reconfiguring the 960 combat-coded PAA into larger squadrons (i.e., increasing the PAA per squadron),[1] (2) adjusting the mix of PAA across the AC and the Reserve Component (RC) of the AFRC and ANG, and (3) adjusting the percentage of the AC PAA assigned to home-station locations in the continental United States (CONUS). Specifically, this research addressed how a change along these three dimensions would affect the Air Force in the following areas:

- Ability to support both surge and steady-state contingency operations
- Ability to absorb the necessary number of F-35 pilots
- Requirements for maintenance manpower and support equipment (SE)
- Requirements for new infrastructure across the set of existing F-16 and A-10 bases[2]
- Ability to develop future senior leaders out of the pool of fighter pilots.

[1] For USAF fighter aircraft, no current squadron has more than 24 PAA. However, fighter squadron sizes have varied over time based on the facilities and aircraft numbers available, and they tend to peak during wartime and decrease during postwar drawdown periods. The analysis presented in this report will examine the potential for squadron sizes larger than 24 PAA to generate increased cost-effectiveness.

[2] We limit the infrastructure analysis to the bases that currently support F-16 and A-10 squadrons, as these are the aircraft the F-35 is designed to replace.

A set of 28 alternative beddowns was examined in this analysis, as presented in Table S.1. These beddowns varied the squadron size between 24 and 36 PAA (for AC and AFRC) and between 18 and 24 PAA (for ANG).[3] These beddowns also varied the percentage of combat-coded PAA in the AC between 45 and 75 percent, and the percentage of total combat-coded AC PAA in CONUS between 50 and 67 percent. The ACC/CC-approved beddown corresponds to beddown alternative 2A.

---

[3] The analyses presented in this report assume that each AFRC and ANG squadron is located at a single base. This assumption is consistent with the current beddown of combat-coded AFRC and ANG fighter/attack squadrons. It is possible that multiple RC squadrons could be assigned to a single wing at a single base, but this analysis did not consider such alternatives. This analysis does, however, examine the efficiencies associated with multisquadron wings for AC squadrons.

Table S.1. F-35 Beddown Alternatives

Percentage of Total PAA in AC	Squadron Size (PAA)	Percentage of Total AC PAA in CONUS	Beddown Alternative	AC Squadrons		RC Squadrons		Total Squadrons
				CONUS	OCONUS	AFRC	ANG	
45	18 (ANG), 24 (AC/AFRC)	50	1A	9	9	7	20	45
		67	1B	12	6	7	20	45
	24 (ANG), 24 (AC/AFRC)	50	1C	9	9	6	16	40
		67	1D	12	6	6	16	40
	18 (ANG), 30 (AC/AFRC)	50	1E	8	7	5	20	40
		67	1F	10	5	5	20	40
	24 (ANG), 30 (AC/AFRC)	50	1G	8	7	5	15	35
		67	1H	10	5	5	15	35
	24 (ANG), 36 (AC/AFRC)	50	1I	6	6	4	16	32
		67	1J	8	4	4	16	32
60	18 (ANG), 24 (AC/AFRC)	50	2A	12	12	4	16	44
		67	2B	16	8	4	16	44
	24 (ANG), 24 (AC/AFRC)	50	2C	12	12	4	12	40
		67	2D	16	8	4	12	40
	18 (ANG), 30 (AC/AFRC)	50	2E	10	9	4	15	38
		67	2F	13	6	4	15	38
	24 (ANG), 30 (AC/AFRC)	50	2G	10	9	5	10	34
		67	2H	13	6	5	10	34
	24 (ANG), 36 (AC/AFRC)	50	2I	8	8	4	10	30
		67	2J	11	5	4	10	30
75	18 (ANG), 24 (AC/AFRC)	50	3A	15	15	4	8	42
		67	3B	20	10	4	8	42
	24 (ANG), 24 (AC/AFRC)	50	3C	15	15	3	7	40
		67	3D	20	10	3	7	40
	18 (ANG), 30 (AC/AFRC)	50	3E	12	12	2	10	36
		67	3F	16	8	2	10	36
	24 (ANG), 30 (AC/AFRC)							
	24 (ANG), 36 (AC/AFRC)	50	3I	10	10	2	7	29
		67	3J	13	7	2	7	29

Key Findings

Potential for Cost Reductions

Our primary finding is that increasing the F-35 squadron size from the levels utilized in the ACC/CC-approved beddown (24 PAA per AC and AFRC squadron, 18 PAA per ANG squadron) can satisfy both expected surge and steady-steady deployment requirements, and can generate significant savings in the following areas:

- Annual pilot absorption flying costs (more than $400 million)
- Annual maintenance manpower costs (more than $180 million)
- One-time support equipment requirements (more than $200 million)
- Annualized facilities costs (more than 10 percent).

The lower bounds on these estimates can be achieved, and all deployment requirements satisfied, were the USAF to implement a posture that utilizes 30 PAA per AC and AFRC squadron and 24 PAA per ANG squadron (beddown alternatives 2G and 2H). The savings would increase were the USAF to select a posture with 36 PAA in AC and AFRC squadrons and 24 PAA in ANG squadrons (alternatives 2I and 2J), but this posture would assume increased risk; it has sufficient squadrons to satisfy surge wartime requirements, but it cannot satisfy steady-state requirements within the desired deploy-to-dwell ratios.

Further savings are possible in all categories except maintenance manpower, if the percentage of PAA in the AC were increased from the level assumed in the ACC/CC-approved beddown (60 percent). The percentage of AC PAA assigned to CONUS locations had little impact on these savings.

Deployment Requirements

We found that all 28 of the beddown alternatives satisfy surge requirements. The surge and steady-state requirements used in this analysis were based upon analysis performed by the Directorate of Studies & Analysis, Assessments and Lessons Learned, Headquarters, U.S. Air Force (AF/A9). These requirements were based on an examination of two of the Integrated Security Constructs (ISCs) developed by the Department of Defense.

A key assumption in this analysis was that each squadron contained one independent or "lead" Unit Type Code (UTC).[4] Thus, each squadron could deploy to and operate out

[4] A UTC is a unit of capability specified by the required manpower and equipment.

of, at most, one location, regardless of squadron size. Figure S.1 demonstrates how the numbers of squadrons available in the 28 alternative F-35 beddowns compare to these squadron requirements. Each marker on the figure corresponds to one paired set of beddown alternatives.[5] The two members of each paired set differ only by the percentage of total AC PAA in CONUS—each member has an equal number of "large" and "small" squadrons. For example, beddowns 1G and 1H each have 20 "large" (in this case, 30 PAA) squadrons and 15 "small" (in this case, 24 PAA) squadrons. The red region on this figure corresponds to the range over which the number of squadrons is insufficient to satisfy the peak surge demand. Observe that all beddown alternatives lie outside of the red region; thus, all have sufficient squadrons to satisfy surge squadron requirements.

Figure S.1. Ability of Alternative F-35 Beddowns to Satisfy Surge Deployment Requirements

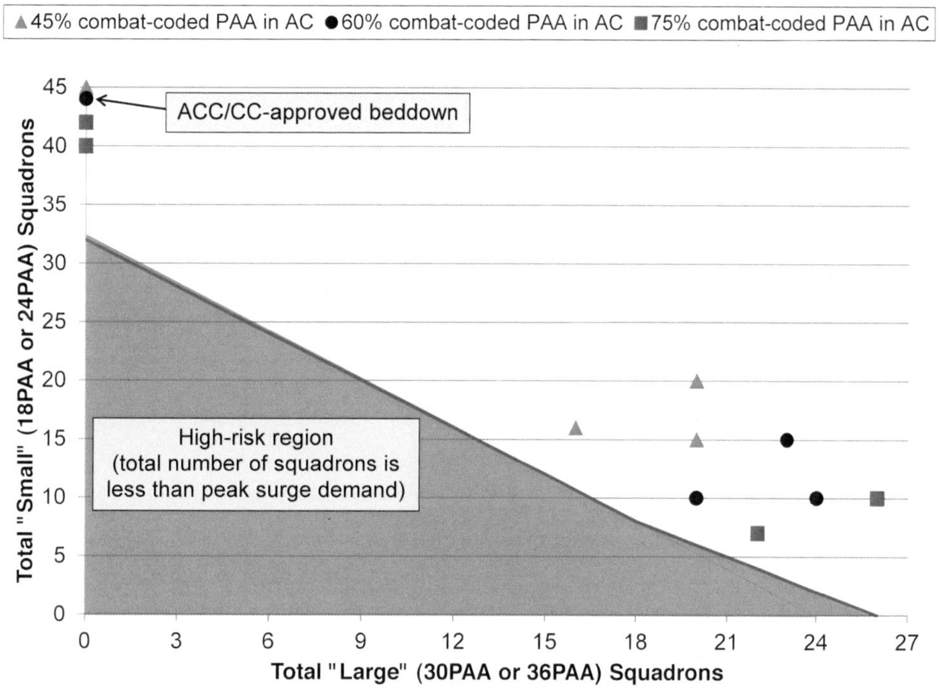

---

[5] The marker at 40 "small" squadrons and zero "large" squadrons actually corresponds to six beddowns: 1C, 1D, 2C, 2D, 3C, and 3D.

Further, most alternatives satisfy rotational requirements within specified deploy-to-dwell ratios. The primary distinction between surge and steady-state rotational requirements is that deploy-to-dwell considerations limit the number of combat-coded squadrons that are available for rotational deployments at any point in time. This analysis assumes that rotational requirements must be satisfied without exceeding the maximum deploy-to-dwell ratios presented in Table S.2. Note that these deploy-to-dwell ratios do not imply any specific deployment duration for any unit; rather, they identify the maximum percentage of time that a unit could be deployed, over an indefinite horizon.

Table S.2. Maximum Allowable Deploy-to-Dwell Ratio for Rotational Requirements

	Non-surge	Post-surge
AC Squadrons	1:3	1:2
RC Squadrons	1:11	1:5

NOTE: The deploy-to-dwell ratios presented in the post-surge column are consistent with current USAF guidance for periods other than surge. This level of deployment is viewed as the maximum supportable level; however, there are concerns that such a high level of deployment poses challenges to the longer-term sustainability of the force. Thus, based upon consultations with ACC, we modified the deploy-to-dwell ratio in non-surge to allow for less deployment stress on the force during non-surge periods. Note that this increases the requirement for the number of squadrons needed during non-surge periods.

This analysis assumes that all RC units and all AC units in CONUS are organized as associate units.[6] It is unclear how this organization into associate units would affect the F-35 force presentation model, and thus the maximum allowable deploy-to-dwell ratio in an RC or AC unit. This analysis assumed that the entire unit is available at the host unit's deploy-to-dwell rate. Alternatively, one could assume that the AC portion of an Active Associate unit was available for deployment at the more-stressing AC rate. However, this poses difficulties from a force presentation concept. If the AC portion is deployed with the rest of its Active Associate unit, force presentation is maintained as an integral squadron. If the AC portion is available at a different rate than the RC portion, then the AC pilots and maintainers would likely need to be sized to support an entire UTC package(s), with separate RC UTCs providing the remainder of a squadron's designed operational capability statement. In this case, the specific UTCs to be supported by the AC portion would need to be identified. Would the AC support an independent ("lead") or dependent ("follow-on") UTC? Would the force presentation of such AC units assume

[6] Thus, every beddown alternative includes both Active Associate units, in which an RC unit has principal responsibility for a weapon system and shares the equipment with an AC unit, and Classic Associate units, in which an AC unit retains principal responsibility for a weapon system and shares the equipment with an RC unit.

that AC UTCs deploy with other AC units, leaving the RC remainder to conduct its home-station mission? Or would AC UTCs deploy with *rainbowed* RC units?[7]

Based upon the required numbers of deployed aircraft and required numbers of deployed locations in the non-surge and post-surge scenarios, we identified the minimum number of squadrons necessary to support rotational requirements.

In Figure S.2, the solid-green region corresponds to the range over which all non-surge and post-surge demands can be satisfied within the deploy-to-dwell ratios identified in Table S.2. The light green region on this figure corresponds to the range over which all rotational requirements can be satisfied within the post-surge deploy-to-dwell ratios. Again, each marker corresponds to one set of paired beddown alternatives that differ only with respect to the percentage of AC PAA in CONUS. For example, beddowns 1I and 1J are represented by a single point on the figure since each has 12 AC squadrons and 20 RC squadrons. Observe that 18 of the 28 beddown alternatives have sufficient squadrons to satisfy rotational requirements within the deploy-to-dwell ratios presented in Table S.1; two additional beddowns can satisfy these requirements if the post-surge deploy-to-dwell ratios were applied during non-surge periods. Both increasing the squadron size (i.e., moving down and to the left within the set of triangles, circles, or squares in the figure) and decreasing the fraction of combat-coded PAA in the AC increase the risk that a beddown alternative will not be able to satisfy rotational deployment requirements within the specified deploy-to-dwell ratios.

It is important to recognize that, under a different employment construct than is currently envisioned in the ISCs (in which the F-35 is deployed in a manner similar to the F-16), the deployment requirements and associated logistics resource requirements might differ significantly from those presented here. Because the employment of the F-35 is still to be determined by the USAF, potential new concepts such as “many locations with very few F-35s at each location” could significantly change these requirements, and thus the supportability of an F-35 beddown that utilizes large squadron sizes.

[7] *Rainbowing* is a deployment strategy used by the RC in which a single deployment requirement is maintained over some duration through the rotation of personnel from multiple RC units.

Figure S.2. Ability of Alternative F-35 Beddowns to Satisfy Rotational Deployment Requirements

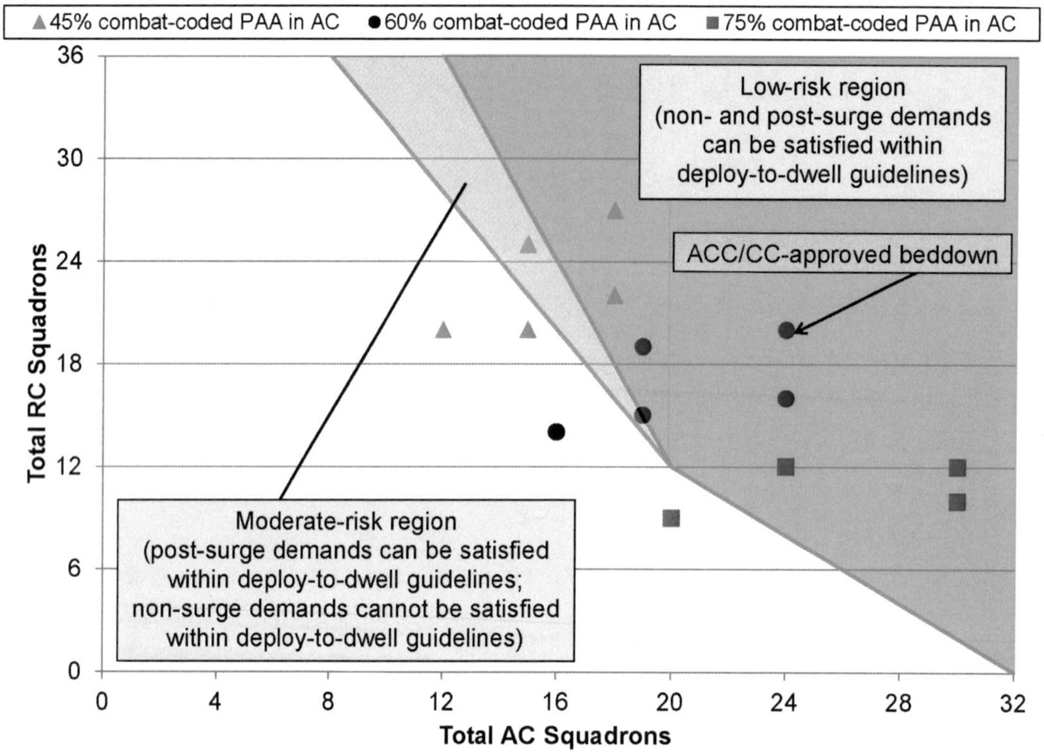

Pilot Absorption

Our analyses of pilot absorption capacities for the beddown options were based on a steady-state absorption model that investigated potential "feasible" absorption conditions,[8] which will create enough experienced pilots to generate adequate pilot inventories, using achievable aircraft utilization (UTE) rates,[9] and maintaining acceptable unit experience levels, while enabling pilots to meet specified minimum Ready Aircrew Program (RAP) training requirements across all units in all components.

The historical norm for fighter pilot absorption has been to fill all AC inventory needs, plus all prior-service Guard and Reserve inventory needs, using pilots absorbed primarily in AC units. This has not been feasible since the late 1990s, however, because

---

[8] Absorption capacities measure the number of new pilots that operational units can absorb per year. Operational units absorb new fighter pilots by providing the training, experience, and supervision needed to develop them into combat pilots, instructors, and leaders. Important factors include the unit manning and experience levels as well as facilities (e.g., simulators, ranges, and airspace) and aircraft utilization rates.

[9] For fighter aircraft, UTE is defined as the average number of sorties flown per PAA per month.

the post–Cold War drawdown took AC force structure below the required levels. For each of the 28 alternative beddowns, we examined three distinct absorption excursions using Active Associations, in which differing numbers of AC pilots operate ANG and AFRC airframes in ANG- and AFRC-assigned units. Similarly, the excursions examined alternatives for Classic Associations (in which ANG and AFRC pilots fly with AC units), which were assumed to exist for every CONUS-based AC unit.

Achieving feasible absorption conditions will require both a change in the burden historically borne by RC units and additional resources allowing AC units to overfly RAP minimums. Only one of the excursions analyzed—the first one—produced pilot inventories that approached the required levels, with a 2.4 to 7.5 percent overfly above the RAP minimums of AC units necessary to satisfy the required inventories. All excursions tended to impose a disproportionate share of the absorption burden on the ANG and AFRC units. The first absorption excursion tested required ANG unit UTE rates that are two to three sorties per PAA per month (15 to 23 percent) greater than the AC UTE for many beddown alternatives, and forced ANG unit experience levels to drop below 60 percent for several beddown alternatives.

We found that squadron size and AC/RC mix affected experience levels in RC units; i.e., RC experience level increases with squadron size and with the percentage of aircraft in the AC. The RC UTE requirement to meet pilot absorption decreases as squadron size increases; this requirement was not significantly affected by the AC/RC mix. Because the number of associated AC pilots per unit does not vary with RC squadron size in the first excursion, the inexperienced AC pilots have a lesser effect on the overall experience level for a larger RC squadron, and the increased flying needed to support the AC pilots is distributed over a larger number of aircraft in larger RC squadrons. As the percent of aircraft in the AC increases, more new AC pilots are absorbed each year, which in turn generates a larger pool of AC pilots who eventually depart the AC as experienced pilots and affiliate with RC units, decreasing the RC units' requirement to train their own inexperienced non-prior-service pilots. The AC UTE requirement decreases as the percentage of total aircraft in AC increases; this requirement was not significantly affected by squadron size. This is because the total AC pilot inventory requirement includes a large number of pilots who are not in F-35 operational units, but who are needed for other missions, such as test and training squadrons, or staff positions. This requirement for AC pilots outside the F-35 operational units was assumed to be constant across all beddown alternatives; thus, alternatives with less aircraft in the AC have fewer AC units through which to absorb the total pilot requirement, whereas alternatives with more aircraft in the AC have a broader base of AC units through which the non-operational units' fighter pilot requirements can be absorbed.

These squadron size effects could have a significant impact on pilot absorption flying costs. Under the first absorption excursion (given the set of UTE requirements identified for each of the 28 alternative beddowns), we identified the annual cost associated with generating the required number of sorties (assuming an average sortie duration of 1.4 flying hours) and a cost of $18,025 per flying hour.[10] Figure S.3 presents the annual pilot absorption costs associated with each of the 28 beddown alternatives (each marker on the figure again corresponds to one paired set of beddown alternatives).

As the fraction of combat-coded PAA in the AC is held constant (i.e., within the set of circles, squares or triangles in the figure), increasing the squadron size (i.e., moving down and to the left on the figure, with fewer squadrons) can significantly reduce the annual pilot absorption flying cost, due to the associated decreased UTE requirements. Observe that the ACC/CC-approved beddown has an annual pilot absorption flying cost of $4.4 billion. Within the alternative that maintains 60 percent of the combat-coded PAA in the AC, increasing AC and AFRC squadron size to 30 PAA while maintaining 18 PAA per ANG squadron can reduce these costs 4 percent relative to the ACC/CC-approved beddown, while increasing ANG squadron size to 24 PAA and maintaining 24 PAA per AC and AFRC squadron could reduce these costs by 8 percent. Increasing both AC and AFRC squadrons to 30 PAA and ANG squadrons to 24 PAA would reduce these costs by 10 percent, while a further increase to 36 PAA in the AC and AFRC could reduce costs by 12 percent.

As the squadron size is held constant, increasing the fraction of combat-coded PAA in the AC (e.g., comparing the marker farthest to the left for each colored set of markers) also generates cost reductions, again due to the associated decreased UTE requirements. When compared to the ACC/CC-approved beddown's $4.4 billion in annual pilot absorption flying costs, there are many alternative beddowns that satisfy all deployment requirements and reduce this cost by 10 percent or more.

---

[10] The Air Force Cost Analysis Agency (AFCAA) provided us with an F-35A steady-state cost per flying hour (CPFH) in base year 2012 dollars. "Steady state" is defined here as the average cost during the period with the maximum number of PAA, which for the F-35A is fiscal years (FYs) 2036–2040. This factor includes cost growth above inflation and comprises costs for fuel ($6,604), consumables ($1,793) and depot level repairables ($9,628).

Figure S.3. Annual Pilot Absorption Flying Costs, by Beddown Alternative

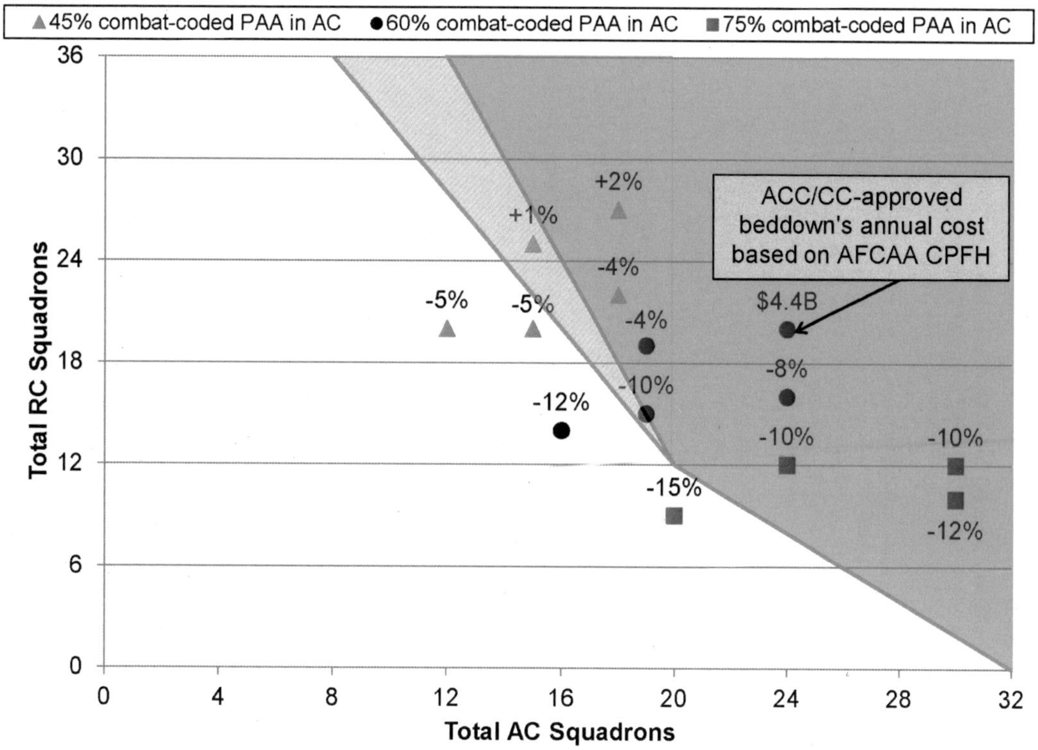

Logistics Resources

We found that increasing squadron size reduces maintenance manpower requirements. For combat-coded aircraft, the required maintenance manpower per PAA decreases as the number of PAA per squadron increases. We estimated that a squadron of 36 PAA could be supported by 26 percent fewer maintenance positions per PAA than could a single squadron of 18 PAA. Furthermore, assigning multiple squadrons to a single wing can generate additional savings beyond those generated by the squadron size effect. Our analysis suggests that a wing of three 36 PAA squadrons requires 6 percent fewer maintenance positions per PAA than a single squadron of 36 PAA.

Figure S.4 presents the total manpower costs associated with each of the 28 beddown alternatives, with the value for each alternative presented as the percentage difference between its cost and the cost of the baseline ACC/CC-approved beddown. Each marker on the figure again corresponds to one paired set of beddown alternatives—each member of the set has an equal number of RC and AC squadrons, they differ only in the percentage of total AC PAA in CONUS.

Figure S.4. Total Annual Maintenance Manpower Costs, by Beddown Alternative

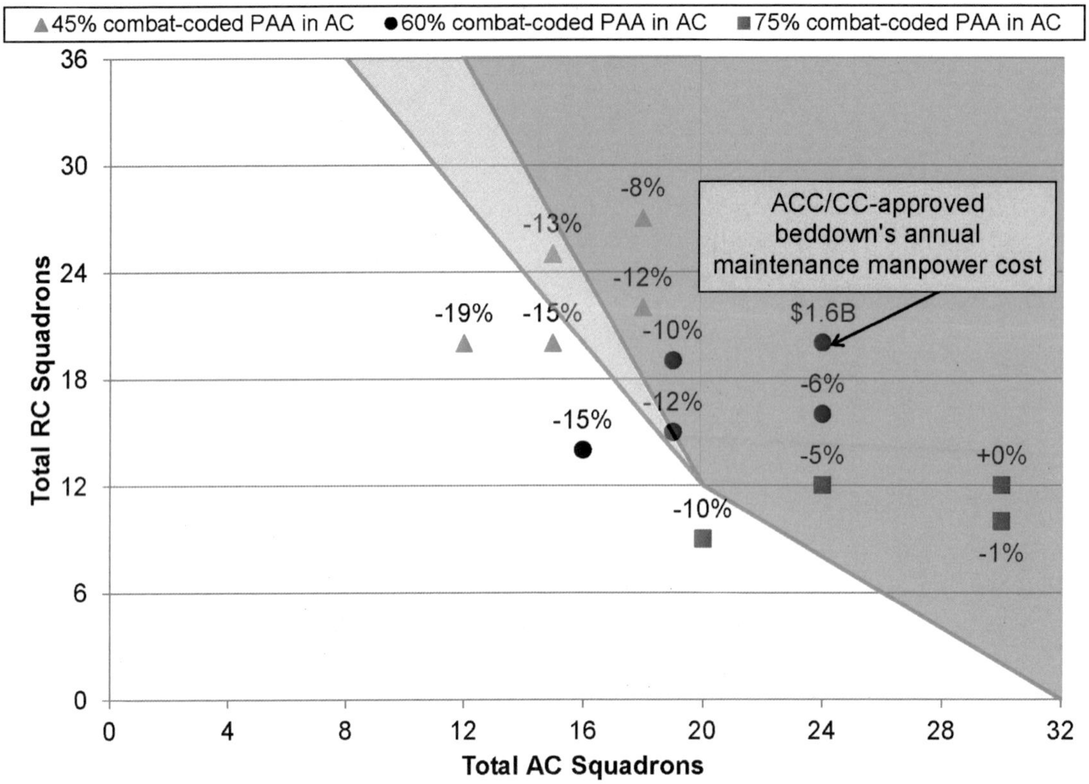

As the fraction of combat-coded PAA in the AC is held constant (i.e., within the set of circles, squares or triangles in the figure), increasing the squadron size (i.e., moving down and to the left on the figure, with fewer squadrons) can significantly reduce the overall maintenance manpower cost. This is consistent with the manpower economies of scale discussed above. However, as the squadron size is held constant, increasing the fraction of combat-coded PAA in the AC (e.g., comparing the marker farthest to the left for each colored set of markers) increases the overall cost. This occurs because the RC is able to make use of part-time maintainers, who are much less expensive in a nondeployed steady-state role than are AC maintainers.

These results assumed that Active Associate units utilize only RC maintenance manpower, with the AC providing no maintenance manpower to the associate units. We also considered two alternative strategies that place AC maintenance manpower in the Active Associate unit. Under the second alternative, the AC would provide the maintenance manpower necessary to support the increased home-station flying caused by

AC pilots.[11] Under the third alternative, in addition to increasing pilot absorption capabilities, the AC manpower is used to generate an increased deployment capability: We identify the AC manpower necessary to support an entire set of UTCs, position these UTCs within Active Associate units, and make these UTCs available at the deploy-to-dwell ratios assumed for the AC. Under both the second and third alternatives, RC full-time manpower is reduced as AC maintenance manpower is added to the unit. However, because the typical AC maintainer would be expected to be less-experienced and less-productive than the typical RC maintainer, these positions were not traded on a one-for-one basis: An equivalency factor of approximately 1.44 AC maintainers per full-time RC maintainer was assumed.

Across the three alternatives considered, the total annual cost differs by no more than 1.3 percent. Said differently, the third alternative provides more deployment capability at essentially the same total cost. However, the number of AC maintainers required at Active Associate units varies significantly. For those beddowns that maintain 60 percent of combat-coded PAA in the AC (as in the ACC/CC-approved beddown 2A), the first and second alternatives can satisfy the AC pilot absorption requirements with between zero and 168 total AC maintenance positions at Active Associate units, while the third alternative provides an increased steady-state deployment capability through the use of between 980 and 1,400 total AC maintenance positions at Active Associate units. Because we found little difference between the manpower composition alternatives with respect to total annual maintenance manpower costs, the key tradeoff to be considered when evaluating these alternatives is the increase in deployment capability that can be achieved under the third alternative versus the increased AC maintenance manpower requirements at Active Associate units.

We found that increasing squadron size reduces SE procurement costs. As the fraction of combat-coded PAA in the AC is held constant, increasing the squadron size can significantly reduce the overall SE procurement cost, because economies of scale also exist for SE requirements. Furthermore, as the squadron size is held constant, increasing the fraction of combat-coded PAA in the AC also decreases the overall cost. This occurs because the ANG is limited to smaller squadron sizes, and when the fraction of total PAA in the AC is increased, fewer PAA are assigned to the smaller ANG squadrons.

---

[11] Note that this increased home-station flying was incorporated into the requirements for the first alternative, but it was not necessary to separate the home-station flying into different segments because RC manpower were performing all maintenance.

Infrastructure

We found that, utilizing current F-16 and A-10 bases, little additional capacity would be required. Our analysis considered infrastructure capacity across six resource categories.[12] As shown in Figure S.5, some of the resource categories proved sufficient for all bases under all beddown alternatives. In particular, for runway and ramp, no new capacity is needed—all F-35 requirements can be satisfied with existing infrastructure. The other resource categories (squadron operations/aircraft maintenance unit (AMU), ammunition storage, corrosion control, and maintenance) did require some additional capacity in most cases (denoted by the cross-hatched areas in the figure). However, these requirements are relatively small.[13]

The beddown alternatives also exhibit some cost reductions associated with consolidation to fewer bases. Larger squadron sizes reduce annualized facilities costs, while increasing the percentage of aircraft in the AC reduces facility costs.

---

[12] This is not an exhaustive list of additional infrastructure required at a current F-16 or A-10 base in order for the base to support F-35 operations. As an example, based on the increased security classification requirements for fifth-generation fighter aircraft, increased cost would be necessary to support a higher level of classification for communications lines, sensitive compartmented information facilities, etc.

[13] Note that this is based on analysis of raw square footage data from an Office of the Secretary of Defense database (the Facility Program Requirements Suite) and does not address the condition or adequacy of current facilities and infrastructure.

Figure S.5. F-35 Infrastructure Requirements, by Beddown Alternative

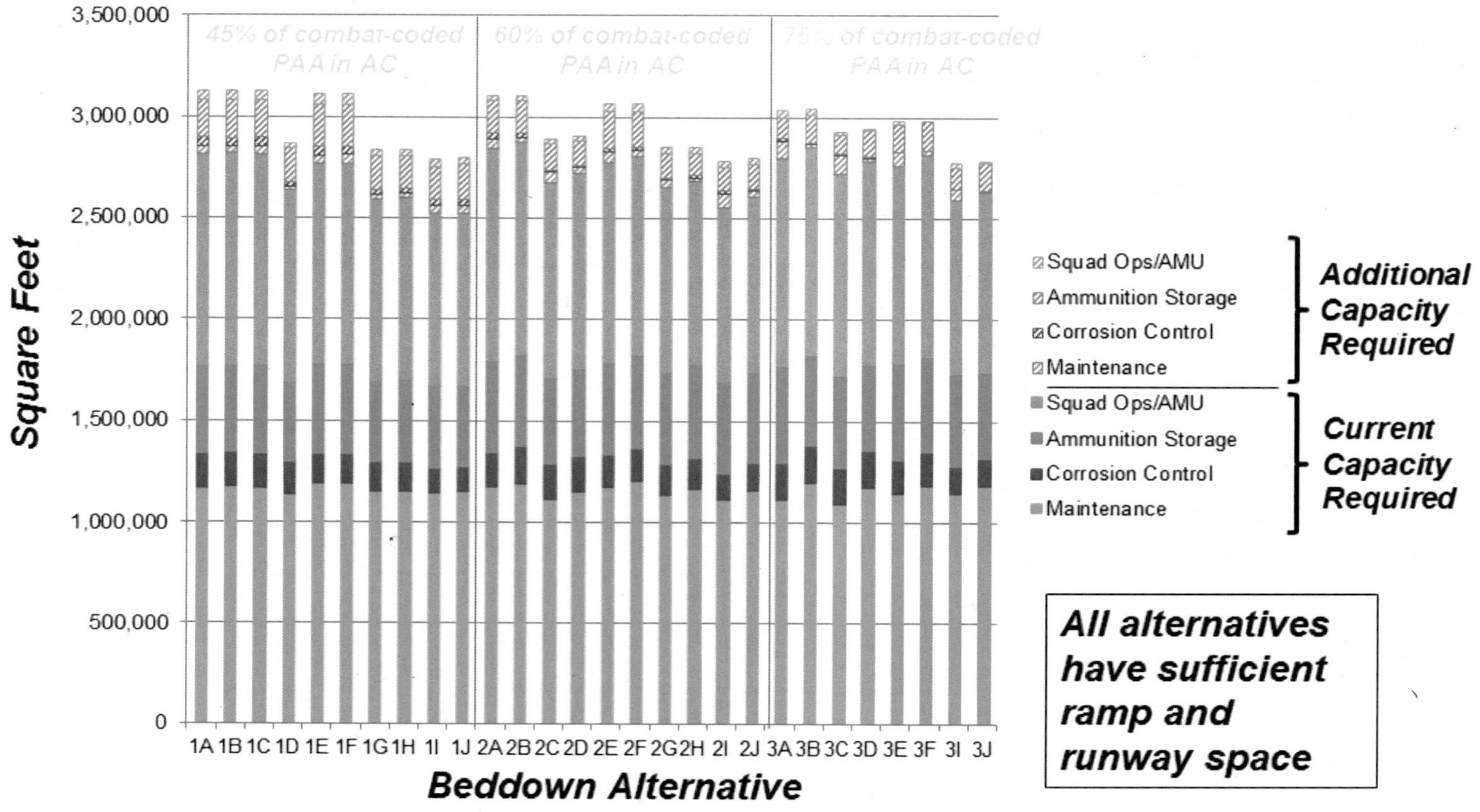

Leader Development

The F-35 beddown alternatives would substantially alter the numbers of PAA and units in the AC, ANG, and AFRC—and, consequently, the numbers of jobs such as squadron commander, group commander, and wing commander that are regarded as key developmental experiences. Hence, Air Force decisionmakers asked whether some alternatives would endanger development of future senior leaders. We assessed the Air Force's capacity for developing fighter pilots in the AC under the beddown alternatives.

Leader development was found to be more affected by the assignment policy used than squadron size or AC/RC mix. All in all, we concluded that the F-35 beddown alternatives would have a slight effect on the AC's capacity for producing future senior leaders with targeted combinations of experience. However, these results suggest that the USAF will be somewhat constrained with respect to fighter pilot leadership development, aside from the impacts of squadron size. To allow for a larger pool of candidates with the preferred characteristics, the USAF needs to be deliberate with its leadership development during the change from legacy fighter/attack aircraft to the F-35, but none of the beddown alternatives with at least 60 percent of the combat-coded PAA in the AC would jeopardize its ability to produce at least as many well-qualified candidates as have actually been promoted to general officer in recent years.

The Way Forward

The findings from this analysis can be used to inform many issues that are within the purview of other USAF analyses and decision processes, including the Total Force Integration Roundtable's discussion of Associate Unit Force Presentation, the Directorate of Strategic Planning, Office of the Deputy Chief of Staff for Strategic Plans and Programs, Headquarters USAF (AF/A8X)'s Multi-Role Fighter Phase II Force Composition Analysis, and the Strategic Basing Process performed by the Office of the Assistant Secretary of the Air Force for Installations, Environment, and Logistics (SAF/IE) and the Office of the Deputy Chief of Staff for Strategic Plans and Programs, Headquarters USAF (AF/A8). In particular, these findings can help determine how F-35 associate units should be composed and resourced in order to meet the requirements of increased pilot absorption and (potentially) increased deployment capability.

Acknowledgments

Many people inside and outside of the Air Force provided valuable assistance and support to our work.[14] We thank Major General Judith Fedder, AF/A4L, the original sponsor of our analysis of maintenance manpower and support equipment, along with Major General John Cooper, AF/A4L, our subsequent sponsor, for their support. We also thank General Philip Breedlove, the Vice Chief of Staff of the Air Force, for requesting that we expand these initial analyses of logistics resources to consider a broader set of USAF resources and requirements; and Major General Brett Williams, Director of Operations, Deputy Chief of Staff for Operations, Plans and Requirements, Headquarters U.S. Air Force (AF/A3O), for his assistance in defining the research directions and briefing our work to the F-35 Executive Review Board.

On their staffs, we extend a special thanks to Colonel Kyle Matyi, Lieutenant Colonel David Seitz, and Guy Fowl from AF/A4L; and to Colonel Jamie Crowhurst, Colonel James Jinnette, Colonel Steven Robinson, Lieutenant Colonel Brian Beales, Lieutenant Colonel Casey Tidgewell, and Major James Caplinger from AF/A3O for all of their assistance.

Staff from the Directorate of Studies and Analyses, Assessments and Lessons Learned, Headquarters U.S. Air Force (AF/A9) were extremely helpful in providing us with deployment requirements for the F-35; in particular we thank David Hickman, Maria Hayden, and Michael Cusack for their assistance to our analysis.

This analysis brought together a Total Force team to ensure that the unique perspectives of the Active Component, Air Force Reserve Command (AFRC) and Air National Guard were represented. We received outstanding support from a very large number of participants at the Air Staff, HQ Air Combat Command, HQ AFRC, and the National Guard Bureau. We greatly appreciate the time and effort that all of the team members spent to provide us with the data—and the background understanding necessary to properly use those data—for our analysis.

At the Joint Program Office, we received assistance and data from Kimberly Fuller, Enass Saad-Pappas, and Sal Baglio. We thank Jennifer Bowles at the Air Force Cost Analysis Agency and Thomas Lies at the Directorate of Cost Analysis, Office of the Deputy Assistant Secretary of the Air Force for Cost and Economics, for providing us with cost data.

[14] All office symbols and military ranks are listed as of the time of this research.

At RAND, we thank our colleagues Laura Baldwin, Manuel Carrillo, Michael McGee, Patrick Mills, Carl Rhodes, and Al Robbert for sharing their insights and suggestions during the course of our analysis. We thank Robert Guffey for his assistance in developing our final project briefing, and are especially grateful to Kristin Leuschner, Megan McKeever, and Jane Siegel for their assistance in the preparation of this document. We thank our RAND colleagues John Ausink, Jeff Hagen and Steve Walters for their thorough reviews; their comments helped shape this report into its final, improved form.

That we received help and insights from those acknowledged above should not be taken to imply that they concur with the findings presented in this report. As always, the analysis and conclusions are solely the responsibility of the authors.

Abbreviations

AC	Active Component
ACA	Aerospace Control Alert
ACC	Air Combat Command
ACC/A4	Directorate of Logistics, Headquarters, Air Combat Command
ACC/A9	Directorate of Studies & Analysis, Assessments and Lessons Learned, Air Combat Command
ACC/CC	commander, Air Combat Command
AFCAA	Air Force Cost Analysis Agency
AF/A4L	Directorate of Logistics, Headquarters, U.S. Air Force
AF/A8X	Directorate of Strategic Planning, Office of the Deputy Chief of Staff for Strategic Plans and Programs, Headquarters USAF
AF/A3O	Directorate of Operations, Deputy Chief of Staff for Operations, Plans and Requirements, Headquarters U.S. Air Force
AFI	Air Force Instruction
AFM	Air Force Manual
AFMA	Air Force Manpower Agency
AFRC	Air Force Reserve Command
AFRCH	Air Force Reserve Command Handbook
AFRC/A4	Logistics Directorate, Headquarters AFRC
AGE	Aerospace Ground Equipment
AMU	Aircraft Maintenance Unit
AMXS	Aircraft Maintenance Squadron
ANG	Air National Guard
ANGH	Air National Guard Handbook
BMC	basic mission capable

C2ISR/EW	Command and Control, Intelligence, Surveillance and Reconnaissance/Electronic Warfare
CAF	combat air forces
CMR	combat mission ready
CONUS	continental United States
CPFH	cost per flying hour
DoD	Department of Defense
E/E	Electronics and Environmental
FY	fiscal year
ISC	Integrated Security Construct
JPO	Joint Program Office
LCOM	Logistics Composite Model
MOS	Maintenance Operations Squadron
MXG	Maintenance Group
MXS	Maintenance Squadron
NDI	Non-Destructive Inspection
NGB	National Guard Bureau
NGB/A1	Manpower Directorate, Headquarters NGB
NGB/A7A	Asset Management Division, Headquarters NGB
O-1—O-6	Officer rankings
O&S	Operating and Support
OCONUS	outside the continental United States
PAA	Primary Aerospace Vehicle Authorized
PAF	Project AIR FORCE
RAP	Ready Aircrew Program
RC	Reserve Component
RIC	Resource Identification Code

SAF/IE	Office of the Assistant Secretary of the Air Force for Installations, Environment, and Logistics
SE	supply equipment
UPT	Undergraduate Pilot Training
USAF	U.S. Air Force
UTC	Unit Type Code
UTE	utilization

1. Introduction

As currently planned, the F-35 Joint Strike Fighter is the largest aircraft acquisition program in Department of Defense (DoD) history. According to the December 2011 F-35 Selected Acquisition Report, the total acquisition cost to procure 2,457 F-35 aircraft across the U.S. Air Force (USAF), Navy, and Marine Corps is $331 billion, with total operating and support (O&S) costs of $617 billion to operate the aircraft through 2065, with both costs computed using a base year of 2012 (DoD, 2011). Moreover, the F-35 cost-per-flying-hour estimate has increased by more than 80 percent (in constant dollars) over the interval 2002 to 2010. To ensure that the affordability of the F-35 program is not threatened by continuing O&S cost growth, the USAF is examining alternative strategies to reduce those costs.

One approach to reducing O&S costs is to reduce the number of F-35 home-station operating locations, which is in turn related to the number of F-35 squadrons and the number of Primary Aerospace Vehicle Authorized (PAA) per squadron. The USAF Program of Record for the F-35A calls for procurement of 1,763 F-35A Conventional Takeoff and Landing-variant aircraft (DoD, 2011). In 2012, the commander of Air Combat Command (ACC/CC) approved a beddown plan to determine how to allocate the 960 combat-coded PAA across fighter squadrons and operating locations. For the purposes of this report, "beddown" refers to the number and sizes of F-35 squadrons, and their distribution across the Active Component (AC), both within the continental United States (CONUS) and outside (OCONUS), the Air Force Reserve Command (AFRC), and the Air National Guard (ANG), without regard to the specific locations at which these squadrons are permanently based.[15] According to the ACC/CC plan, there are to be 44 squadrons, with the aircraft allocated into squadrons of 24 PAA in the AC and AFRC, and squadrons of 18 PAA in the ANG. These 44 total squadrons would be distributed among 31 operating locations. The remaining 803 noncombat-coded aircraft would be assigned to other missions, such as training or testing (369 total aircraft), or would fill requirements for backup aircraft inventory or attrition reserve (434 total aircraft).

At the request of the Vice Chief of Staff of the Air Force, RAND Project AIR FORCE (PAF) assessed whether O&S savings could be achieved (1) by reconfiguring the

[15] We recognize that this use of the term "beddown" is inconsistent with AFI 10-503 (2010), which states, "Beddown is considered the execution of a basing action." We use the term "beddown" here in a different manner, and avoid the use of the term "basing", to emphasize that this analysis is not focused on specific locations for permanently stationed F-35 units.

960 combat-coded PAA into larger squadrons (i.e., increasing the PAA per squadron),[16] (2) by adjusting the mix of PAA across the AC and Reserve Component (RC), and (3) by adjusting the percentage of the AC PAA assigned to home-station locations in CONUS.

Within this analysis, we limited our focus to combat-coded aircraft. We excluded backup inventory and attrition reserve aircraft from the analysis because these aircraft do not generate significant O&S costs, relative to combat-coded, test, and training PAA. We did not include training and test PAA in this analysis because these aircraft were already relatively concentrated in the ACC/CC beddown, suggesting that relatively small efficiencies could be gained through further consolidation.

Specifically, this research addressed how a change along these three dimensions would affect the Air Force in the following ways:

- Ability to support both surge and steady-state contingency operations
- Ability to absorb the necessary number of F-35 pilots
- Requirements for maintenance manpower and support equipment (SE)
- Requirements for new infrastructure across the set of existing F-16 and A-10 bases
- Ability to develop future senior leaders out of the pool of fighter pilots.

A key tenet of this analysis is that it was not intended to make specific recommendations about the utility of any specific site as a potential F-35 beddown location. Instead, this analysis will focus on issues that are generally not site-specific.[17]

This analysis addresses aspects of many issues that are within the purview of other USAF analyses and decision processes, namely the Force Composition Analysis performed by the Directorate of Strategic Planning, Office of the Deputy Chief of Staff for Strategic Plans and Programs, Headquarters USAF (AF/A8X) and the Strategic Basing Process performed by the Office of the Assistant Secretary of the Air Force for Installations, Environment, and Logistics (SAF/IE) and the Office of the Deputy Chief of Staff for Strategic Plans and Programs, Headquarters USAF (AF/A8). The analysis presented in this report is not intended to be duplicative, nor is it an attempt to validate the findings of these other efforts. Rather, the findings of this analysis should be useful to inform those (and other) efforts that are examining similar issues related to the USAF F-35 beddown.

---

[16] For USAF fighter aircraft, no current squadron has more than 24 PAA. However, fighter squadron sizes have varied over time based on the facilities and aircraft numbers available, and they tend to peak during wartime and decrease during postwar drawdown periods. The analysis presented in this report will examine the potential for squadron sizes larger than 24 PAA to generate increased cost-effectiveness.

[17] The infrastructure requirements analysis is an exception to this rule, but the findings presented in this section will not be focused at the level of individual locations.

Current F-35 Beddown Plan

In 2012, ACC/CC approved a beddown plan to identify the allocation of the 960 combat-coded PAA across fighter squadrons and operating locations. This plan identifies the following sets of F-35 operating locations for combat-coded aircraft:

- AC, within CONUS: three squadrons of 24 PAA at each of three locations; two squadrons of 24 PAA at one location; one squadron of 24 PAA at one location (for a total of five locations, 12 squadrons, and 288 PAA)
- AC, OCONUS: two squadrons of 24 PAA at each of six locations (for a total of six locations, 12 squadrons, and 288 PAA)
- AFRC: one squadron of 24 PAA at each of four locations (for a total of four locations, four squadrons, and 96 PAA)
- ANG: one squadron of 18 PAA at each of 16 locations (for a total of 16 locations, 16 squadrons, and 288 PAA)

The sum totals across all combat-coded F-35s are 31 locations, 44 squadrons, and 960 PAA.

F-35 Beddown Alternatives

Within this analysis, we will consider a set of 28 F-35 beddown alternatives. These alternatives vary across three dimensions. First, we considered three values for the percentage of total combat-coded F-35 PAA in the AC: 45 percent, 60 percent and 75 percent. Second, this analysis considered three values for the number of PAA per squadron in the AC: 24 PAA, 30 PAA and 36 PAA. Based upon consultation with the National Guard Bureau (NGB) and Headquarters AFRC, it was determined that within any beddown alternative, AFRC squadrons would always be assumed to have the same number of PAA per squadron as AC squadrons. However, only two values would be considered for the number of PAA per ANG squadron: 18 PAA and 24 PAA. Third, we considered two values for the percentage of AC PAA that would be based at CONUS home-station locations: 50 percent and 67 percent.[18] In keeping with the ACC/CC-approved plan's allocation across the RC, all beddowns that were considered place significantly more PAA in the ANG than in the AFRC. Table 1.1 presents the full set of

[18] The values presented here for "Percentage of Total PAA in AC" and "Percentage of Total AC PAA in CONUS" are approximate, as it is not always possible to apportion the PAA in squadrons of the identified size and also obtain the exact percentages specified in the AC and AC CONUS.

alternative F-35 beddowns considered in this analysis, and introduces the naming convention that will be used throughout the remainder of this report.[19]

Observe that the ACC/CC-approved beddown corresponds to beddown alternative 2A. Throughout this report, we will examine the impact of these 28 beddown alternatives on each of the five analytic focus areas (contingency requirements, pilot absorption, logistics requirements, infrastructure requirements and leadership development), with respect to the performance of the ACC/CC-approved beddown.

More detail regarding how this number of squadrons was arranged into multisquadron wings for the AC is presented in Table 1.2. We will assume that each AFRC and ANG squadron is located at a unique base.[20] We assume that each AC Wing, as presented in Table 1.2, corresponds to one base. For example, beddown 1A corresponds to 7 AFRC bases; 20 ANG bases; three AC CONUS bases, with three squadrons of 24 PAA at each base; and four AC OCONUS bases, one base with three squadrons of 24 PAA, and two squadrons of 24 PAA at each of the other three AC OCONUS bases.[21]

---

[19] Note that, under this structure, there are two beddown alternatives (3G and 3H) that are not included in this analysis. We exclude them due to the difficulty of allocating 25 percent of the total combat-coded PAA to the RC with AFRC squadrons of 30 PAA and ANG squadrons of 24 PAA. The only option for which the arithmetic works has a total of 120 PAA in AFRC and 120 PAA in ANG, which is inconsistent with all other beddowns, for which ANG has many more PAA than does AFRC.

[20] This assumption is consistent with the current beddown of combat-coded AFRC and ANG fighter/attack squadrons. It is possible that multiple RC squadrons could be assigned to a single wing; however this analysis did not consider such alternatives.

[21] There are significant cost implications associated with changing the number of USAF bases at which fighter/attack aircraft are permanently stationed. Issues related to the closure or repurposing of existing USAF bases were beyond the scope of this analysis.

Table 1.1. F-35 Beddown Alternatives

Percentage of Total PAA in AC	Squadron Size (PAA)	Percentage of Total AC PAA in CONUS	Beddown Alternative	AC Squadrons		RC Squadrons		Total Squadrons
				CONUS	OCONUS	AFRC	ANG	
45	18 (ANG), 24 (AC/AFRC)	50	1A	9	9	7	20	45
		67	1B	12	6	7	20	45
	24 (ANG), 24 (AC/AFRC)	50	1C	9	9	6	16	40
		67	1D	12	6	6	16	40
	18 (ANG), 30 (AC/AFRC)	50	1E	8	7	5	20	40
		67	1F	10	5	5	20	40
	24 (ANG), 30 (AC/AFRC)	50	1G	8	7	5	15	35
		67	1H	10	5	5	15	35
	24 (ANG), 36 (AC/AFRC)	50	1I	6	6	4	16	32
		67	1J	8	4	4	16	32
60	18 (ANG), 24 (AC/AFRC)	50	2A	12	12	4	16	44
		67	2B	16	8	4	16	44
	24 (ANG), 24 (AC/AFRC)	50	2C	12	12	4	12	40
		67	2D	16	8	4	12	40
	18 (ANG), 30 (AC/AFRC)	50	2E	10	9	4	15	38
		67	2F	13	6	4	15	38
	24 (ANG), 30 (AC/AFRC)	50	2G	10	9	5	10	34
		67	2H	13	6	5	10	34
	24 (ANG), 36 (AC/AFRC)	50	2I	8	8	4	10	30
		67	2J	11	5	4	10	30

Percentage of Total PAA in AC	Squadron Size (PAA)	Percentage of Total AC PAA in CONUS	Beddown Alternative	AC Squadrons		RC Squadrons		Total Squadrons
				CONUS	OCONUS	AFRC	ANG	
75	18 (ANG), 24 (AC/AFRC)	50	3A	15	15	4	8	42
		67	3B	20	10	4	8	42
	24 (ANG), 24 (AC/AFRC)	50	3C	15	15	3	7	40
		67	3D	20	10	3	7	40
	18 (ANG), 30 (AC/AFRC)	50	3E	12	12	2	10	36
		67	3F	16	8	2	10	36
	24 (ANG), 30 (AC/AFRC)							
	24 (ANG), 36 (AC/AFRC)	50	3I	10	10	2	7	29
		67	3J	13	7	2	7	29

Table 1.2. Arrangement of AC Squadrons Into Multisquadron Wings, for Each Alternative Beddown

		Number of Wings with *N* Squadrons						
		AC CONUS				AC OCONUS		
Beddown Alternative	**AC Squadron Size (PAA)**	**1**	**2**	**3**	**4**	**1**	**2**	**3**
1A	24			3			3	1
1B				4		2	2	
1C				3			3	1
1D				4		2	2	
1E	30		1	2		1	3	
1F			2	2		3	1	
1G			1	2		1	3	
1H			2	2		3	1	
1I	36			2		2	2	
1J			1	2		4		
2A	24	1	1	3			6	
2B			2	4			4	
2C				4				4
2D			2	4			4	
2E	30		2	2			3	1
2F			2	3		2	2	
2G			2	2			3	1
2H			2	3		2	2	
2I	36		1	2			4	
2J			1	3		3	1	

		Number of Wings with *N* Squadrons						
		AC CONUS				AC OCONUS		
Beddown Alternative	**AC Squadron Size (PAA)**	**1**	**2**	**3**	**4**	**1**	**2**	**3**
3A	24			1	3			5
3B					5		2	2
3C				1	3			5
3D					5		2	2
3E	30			4				4
3F			2	4			4	
3I	36		2	2			2	2
3J			2	3		1	3	

An important aspect of this analysis is that it was directed to assume that all RC units and all AC units in CONUS would utilize associate unit arrangements. Thus, every beddown alternative will include both *Active Associate* units, in which an RC unit has principal responsibility for a weapon system and shares the equipment with an AC unit, and *Classic Associate* units, in which an AC unit retains principal responsibility for a weapon system and shares the equipment with an RC unit.

One of the principal motivations for the use of Active Associate units in the recent past has been to increase pilot absorption by assigning inexperienced AC pilots to Active Associate units and using the highly experienced RC pilot force to relieve some of the AC pilot training burden.[22] Active Associate units have also been viewed as a potential means for increasing access to RC force structure for deployments during steady-state periods, since the AC pilots and maintainers assigned to an Active Associate unit could be available for deployment more often than RC personnel.[23]

Guidance is required to identify whether Active Associate units for the F-35 are intended to achieve both, or only the former, of these objectives. This decision affects the force presentation concept, and thus deployment capability, as well as costs of these units. These issues will be discussed in more detail in later sections of this report.

Organization of This Report

The remainder of this report is organized into six chapters. Chapter Two presents our analysis of surge and steady-state contingency requirements for the F-35. Chapter Three describes our analysis of pilot absorption requirements and capabilities. Chapter Four contains our analysis of logistics resource requirements. Chapter Five presents our analysis of F-35 infrastructure requirements, in relation to the set of existing F-16 and A-10 bases. Chapter Six describes our analysis of impacts on leadership development. Finally, Chapter Seven contains our conclusions.

---

[22] By similar logic, Active Associate units could also assist with the training of AC maintenance personnel (although this would force RC maintainers to spend less time on direct production tasks and more time supporting maintenance training, eliminating some of the efficiency advantage that has traditionally been enjoyed by RC maintenance manpower).

[23] The motivation for Classic Associate units is different. By assigning experienced RC pilots to the unit, the unit can reduce its total costs and increase its overall experience level. However, Classic Associate units do not significantly increase pilot absorption capabilities, and could actually reduce access to AC PAA during steady-state periods.

2. Deployment Requirements

In this chapter, we present the results of our analysis of the relationship between F-35 squadron size (i.e., PAA per squadron) and the ability of the combat-coded F-35 fleet to satisfy surge and steady-state contingency requirements.

A key assumption that was made in this analysis was that each squadron contained one independent or "lead" Unit Type Code (UTC).[24] Thus, each squadron could deploy to and operate out of at most one location, regardless of squadron size, consistent with USAF policy for resourcing legacy fighter squadrons for deployment. Said differently, this assumption implies that if 72 PAA were organized into three squadrons of 24 PAA each, then these 72 PAA could deploy to and operate out of no more than three locations. If, instead, these 72 PAA were organized into two squadrons of 36 PAA each, then these 72 PAA could deploy to and operate out of no more than two locations. If the fleet size remains constant, larger squadron sizes reduce the overall number of F-35 squadrons, thereby reducing the number of lead UTCs and thus the number of locations to which the USAF can simultaneously deploy and conduct F-35 operations.

The surge and steady-state requirements used in this analysis were based upon analysis performed by the Directorate of Studies & Analysis, Assessments and Lessons Learned, Headquarters, U.S. Air Force (AF/A9). These requirements were based on an examination of two of the Integrated Security Construct (ISC) scenarios developed by the Department of Defense.

This analysis examined the impact of squadron size on the fleet of all 960 combat-coded F-35s in the USAF's Program of Record. According to the Joint Program Office (JPO) 2011 USAF F-35A Beddown Plan,[25] 2034 will be the earliest year that the USAF will possess all 960 combat-coded F-35. Although the ISC scenarios do not correspond to this 2034 timeframe, we utilized them nonetheless, making two modifications.

According to the acquisition report, the F-35A is intended to "replace the F-16 and A-10 and complement the F-22" (DoD, 2011). Thus, we assumed that as of 2034, there are no combat-coded A-10 or F-16 available for tasking. On every occasion that the ISC scenarios contain a demand for an F-16 or A-10, we replaced that demand with an F-35 on a one-aircraft-for-one aircraft basis.[26]

---

[24] A UTC is a unit of capability specified by the required manpower and equipment.

[25] Personal correspondence.

[26] Note that we did not assume that F-35s would replace F-15 or F-22 demands in the ISCs.

The second modification reflects the impact of squadron size on deployment beddowns. The number of aircraft deployed to a location during contingency operations is influenced by the assumed squadron size for that aircraft. For an aircraft such as the F-16, which is currently organized primarily into combat-coded squadrons of 18 PAA and 24 PAA, most planning scenarios call for the number of aircraft deployed to any individual site to be a value that is the sum of an integer multiple of 18 and an integer multiple of 24 (e.g., 60 aircraft, which is equal to two squadrons of 18 PAA each and one squadron of 24 PAA). Were the USAF to modify the F-35 squadron size to some other value, such as 30 PAA, it is reasonable to assume that the planning scenarios would have their deployment beddowns modified to reflect the new squadron size. Thus, for alternative F-35 beddowns that utilized a squadron size other than 18 PAA or 24 PAA, we modified the ISC scenario beddowns to be reflective of the new squadron size, with two constraints: (1) no changes were made to the total number of PAA deployed for any scenario; and (2) the same number of deployment locations were maintained for each scenario. Thus, if a given ISC scenario called for 120 F-35s to be deployed across four operating locations, our modified scenarios may have changed the specific numbers of aircraft at each location, but our modified scenarios would maintain a total of 120 F-35s deployed across four operating locations.

Each ISC has a non-surge, surge, and post-surge period. There are four types of ISC activities that can potentially require F-35 support: campaign (i.e., warfight), Aerospace Control Alert (ACA), rotational foundational activities and non-rotational foundational activities.[27]

Surge Deployment Requirements

Let us first consider the surge requirements. For surge, we will assume that all combat-coded squadrons in both the AC and RC are available for tasking. When a unit is performing ACA, it maintains fighter aircraft in ready state to defend domestic U.S. airspace. For ACA support during surge, we will assume that any individual squadron can be assigned an ACA mission and can support ACA operations at two simultaneous locations, operating from a deployed location in the United States and from home station. A squadron assigned to an ACA mission is utilizing its lead UTC for ACA support, and is thus not available for further deployment to support campaign activities.

---

[27] Foundational activities are the steady-state operations conducted by both permanently stationed and regularly deployed forces across the globe providing deterrence and performing episodic and ongoing activities that represent day-to-day demands. Foundational activities also include strategic deterrence and homeland defense. Non-rotational foundational activities are the "presence" sort of activities that require squadrons in permanent locations, e.g., South Korea, Japan and Europe. Rotational foundational activities include activities similar to recent support in U.S. Central Command, U.S. Africa Command, etc.

For campaign activities, the squadron size has an impact on the number of squadrons that need to be deployed. This is because, in surge scenarios, it is not uncommon for more than one squadron's worth of aircraft to be deployed to one location. Based on the modified ISC scenarios, we identified the number of squadrons that would need to deploy to satisfy surge requirements. We present these requirements in two forms. Each alternative beddown identified in Chapter One has an associated squadron size for AC and AFRC units, and another (potentially different) squadron size for ANG units. Based upon the number of deployed aircraft and number of deployed locations in the surge scenarios, we identified the minimum number of squadrons necessary to support the surge requirements. Out of this minimum number of squadrons, some would need to be of the larger squadron size associated with the AC and AFRC, while some could potentially be supported with a smaller squadron from the ANG. As an example, if the F-35 beddown is assumed to utilize squadrons of 30 PAA and 18 PAA, then a deployed location with 45 PAA could be supported with a minimum of two squadrons, one of which must have 30 PAA, the second of which could potentially have 18 PAA. Table 2.1 presents these surge requirements for the ISC posing the greatest demand on F-35 deployment.

Table 2.1. Surge Deployment Requirements

Squadron Size Mix	18 PAA and 24 PAA	18 PAA and 30 PAA	All 24 PAA	24 PAA and 30 PAA	24 PAA and 36 PAA
Minimum total squadrons required	32	26	32	26	24
Must be larger squadron size	19	19	32	18	16
Could be smaller squadron size	13	7	n.a.	8	8

Figure 2.1 demonstrates how the number of squadrons available in the set of 28 alternative F-35 beddowns compares to the squadron requirements shown in Table 2.1. The red region in the figure corresponds to the range over which the number of squadrons is insufficient to satisfy the peak surge demand. It is bounded by the line that intersects the vertical axis at 32 (the minimum total squadrons required when all squadrons are 18 PAA or 24 PAA, according to Table 2.1) and the horizontal axis at 26 (the minimum total squadrons required when squadrons of 30 or 36 PAA are used, according to Table 2.1). Each marker on the figure corresponds to one paired set of beddown alternatives.[28] The two members of each paired set differ only by the percentage of total AC PAA in CONUS—each member has an equal number of "large" and "small" squadrons. For example, beddowns 1G and 1H each have 20 "large" squadrons (in this case, 30 PAA) and 15 "small" ones (in this case, 24 PAA). Observe that all beddown alternatives lie outside the red region on this figure, meaning **all 28 alternatives have sufficient squadrons to satisfy surge squadron requirements.**

[28] The marker at 40 "small" squadrons and zero "large" squadrons actually corresponds to six beddowns: 1C, 1D, 2C, 2D, 3C, and 3D.

Figure 2.1. Ability of Alternative F-35 Beddowns to Satisfy Surge Deployment Requirements

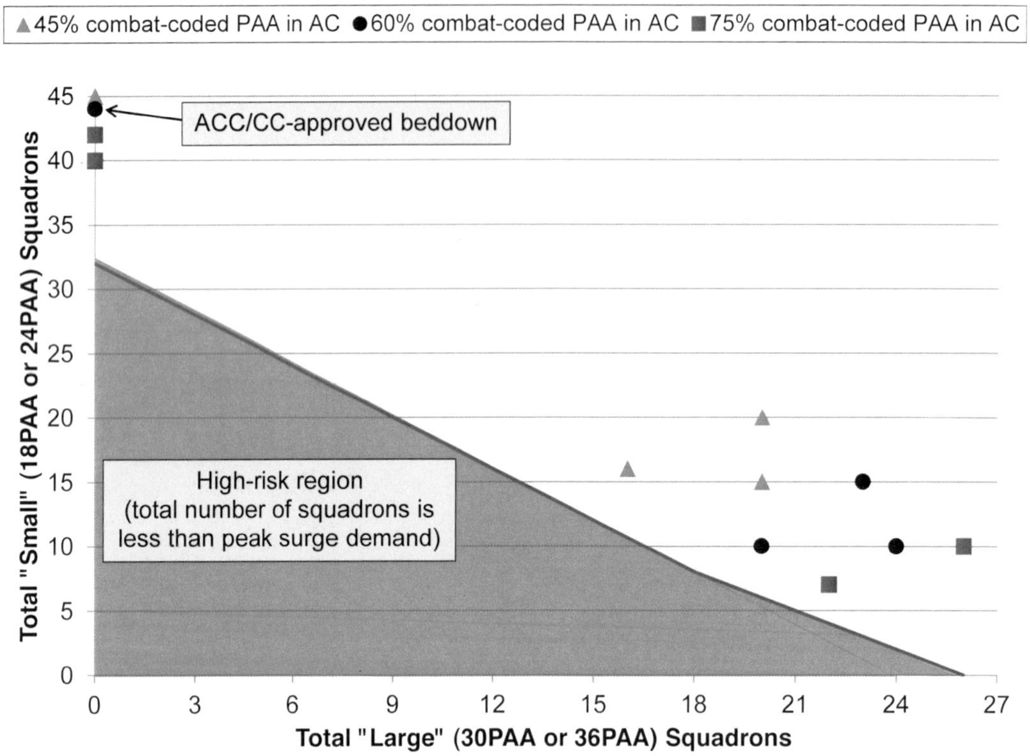

Rotational Deployment Requirements

The primary distinction between surge and steady-state rotational requirements is that for rotational requirements, deploy-to-dwell considerations limit the number of combat-coded squadrons that are available for deployment at any point in time. Note that these deploy-to-dwell ratios do not imply any specific deployment duration for any unit; rather they identify the maximum percentage of time that a unit could be deployed, over an indefinite horizon. A deploy-to-dwell ratio of $1{:}n$ suggests that, out of every $n+1$ time periods, a single squadron can be deployed no more than one period. Thus, $n+1$ total squadrons are needed in order to maintain a permanently deployed presence of one squadron at one location without violating a maximum deploy-to-dwell ratio of $1{:}n$ for any squadron. This analysis assumes that rotational foundational activities require the deployment of combat-coded squadrons, in accordance with the maximum deploy-to-dwell ratios presented in Table 2.2.

Table 2.2. Maximum Allowable Deploy-to-Dwell Ratio for Rotational Requirements

	Non-surge	Post-surge
AC squadrons	1:3	1:2
RC squadrons	1:11	1:5

NOTE: The deploy-to-dwell ratios presented in the post-surge column are consistent with current USAF guidance for periods other than surge. This level of deployment is viewed as the maximum supportable level; however, there are concerns that such a high level of deployment poses challenges to the longer-term sustainability of the force. Thus, based upon consultations with ACC, we modified the deploy-to-dwell ratio in non-surge to allow for less deployment stress on the force during non-surge periods. Note that this increases the requirement for the number of squadrons needed during non-surge periods.

Force Presentation Options for Associate Units Impact Deploy-to-Dwell Constraints

This analysis assumes that all RC units and all AC units in CONUS are organized as associate units. It is unclear how this organization into associate units would affect the force presentation model, whereby USAF forces are deployed and employed in contingency operations, and thus the maximum allowable deploy-to-dwell ratio in an RC or AC unit. This analysis assumed that the entire unit is available at the host unit's deploy-to-dwell rate. Thus, we assumed that the AC portion of an Active Associate unit was available for rotational deployment at the RC rate. We assumed that the RC portion of a Classic Associate unit was available at the AC rate. This is a conservative assumption for Active Associate units, since it restricts access to the Active Associate AC forces to the lesser availability of RC units. This assumption could be problematic for Classic Associate units, but in the specific sets of deployment requirements that were examined in this analysis, the non-surge and post-surge deployments typically required less than a full squadron's worth of aircraft and could thus be supported with the AC portion of a Classic Associate unit, provided that the RC portion of such units was not very large.[29]

Alternatively, one could assume that the AC portion of an Active Associate unit was available for deployment at the AC rate. However, this poses difficulties from a force presentation concept. If the AC portion is deployed with the rest of its Active Associate unit, force presentation is maintained as an integral squadron. If the AC portion is available at a different rate than the RC portion, then the AC pilots and maintainers

[29] In Chapter Three we will discuss alternatives for the number of RC pilots that are included in a Classic Associate unit.

would likely need to be sized to support an entire UTC package(s), with separate RC UTCs providing the remainder of a squadron's designed operational capability statement. In this case, the specific UTCs to be supported by the AC portion would need to be identified. Would the AC support an independent ("lead") or dependent ("follow-on") UTC? Would the force presentation of such AC units assume that AC UTCs deploy with other AC units, leaving the RC remainder to conduct its home-station mission? Or would AC UTCs deploy with *rainbowed* RC units?[30] These force presentation issues were beyond the scope of this analysis, but will require additional study should the USAF decide to utilize associate unit arrangements in the majority of its F-35 squadrons. Although we will not examine the impact of different deploy-to-dwell ratios for the AC and RC portions of an Active Associate on the rotational deployment requirements in this analysis, Chapter Four will examine the related impact on maintenance manpower requirements.

Evaluating an Alternative Beddown's Ability to Satisfy Rotational Requirements

During non-surge and post-surge periods, we assumed that ACA support could be performed at home stations by nondeployed units, and thus would not consume any lead UTCs. Non-rotational foundational activities were assumed to be performed by AC units at permanent OCONUS beddown locations; the number of squadrons required to perform non-rotational foundational activities was subtracted from the pool of AC squadrons available to support rotational deployments. This analysis did not address the impact on noncontingency, non-rotational demands, such as Red Flag exercises, during a squadron's nondeployed (i.e., "dwell") period, since these demands are often difficult to quantify and are not viewed as requirements for force-sizing analysis.

Based upon the number of deployed aircraft and number of deployed locations in the non-surge and post-surge scenarios, we identified the minimum number of squadrons necessary to support rotational requirements. These rotational requirements vary slightly based on the squadron size, but the effect is much less than that observed for surge requirements. This is because most non-surge and post-surge operating locations require a relatively small number of aircraft, typically less than the squadron sizes under consideration. Based on the assumptions of this analysis, a notional operating location that requires 12 F-35s will require one squadron, whether the assumed squadron size is 18 PAA or 36 PAA. We will assume that such a deployment counts as a deployment period for the entire squadron; which is a conservative estimate since not all members of

---

30 *Rainbowing* is a deployment strategy used by the RC in which a single deployment requirement is maintained over some duration through the rotation of personnel from multiple RC units.

the squadron would be need to be deployed, and thus the average deploy-to-dwell ratio for the individuals in such a squadron would be less than what we compute here.

We found that the maximum total number of squadrons that must be deployed at any point in time during non-surge and post-surge periods is less than the total number of squadrons that must be deployed during surge periods, thus all alternatives have a sufficient number of squadrons to satisfy all non-surge and post-surge demands. What is less apparent is if there are sufficient squadrons to satisfy these rotational requirements within the specified deploy-to-dwell ratios.

Post-surge and non-surge squadron requirements are particularly sensitive to the AC/RC mix. Figure 2.2 demonstrates how the number of squadrons available in the set of 28 alternative F-35 beddowns compares to these squadron requirements. Again, each marker on the figure corresponds to one paired set of beddown alternatives. For example, beddowns 1I and 1J each have 12 AC squadrons and 20 RC squadrons. We base all non-surge and post-surge demands on the requirements associated with a beddown alternative that utilizes all squadrons of 24 PAA.[31] The solid-green region on this figure corresponds to the range over which all non-surge and post-surge demands can be satisfied within the deploy-to-dwell ratios identified in Table 2.2. The light green region on this figure corresponds to the range over which all rotational requirements can be satisfied within the post-surge deploy-to-dwell ratios. The point at which the green line intersects the horizontal axis indicates that an F-35 posture consisting of 32 AC squadrons and zero RC squadrons would be sufficient to satisfy rotational requirements within the specified deploy-to-dwell constraints. The point at which the green line farthest to the left intersects the top of the figure indicates the minimum number of AC squadrons (eight) and the corresponding number of RC squadrons (36) that would be necessary to satisfy all rotational requirements within the post-surge deploy-to-dwell guidelines.[32] Observe that most beddown alternatives lie within the solid-green region on this figure. Of the 28 alternatives, 18 have sufficient squadrons to satisfy rotational requirements within the specified deploy-to-dwell ratios, and two additional beddowns could satisfy these requirements if the post-surge deploy-to-dwell ratios were applied during non-surge periods. For beddowns with 75 percent or 60 percent of the combat-coded PAA in the AC, all alternatives except for those with 36 PAA in AC and AFRC squadrons were able

---

[31] That is, in order to graph the darkand light green regions as a function of only the number of RC and AC squadrons, one needs to specify the required number of deployed squadrons. For Figure 2.2, we use the required number of deployed squadrons, assuming all squadrons are of 24 PAA. While slightly fewer deployed squadrons are required for alternatives with 24 PAA squadrons in ANG and 30 or 36 PAA squadrons in AC and AFRC, these reductions are not so large as to bring alternatives that are shown as lying outside the darkor light green regions on Figure 2.2 back inside these regions.

[32] Although it is not shown on this graphic, if all rotational requirements needed to be satisfied within the deploy-to-dwell ratios presented in Table 2.2, then a minimum of eight AC squadrons would again be required, but the corresponding RC requirement would now be 48 squadrons.

to satisfy rotational requirements within the specified deploy-to-dwell ratios. For beddowns with 45 percent of the combat-coded PAA in the AC, only alternatives with 24 PAA in AC and AFRC squadrons were able to satisfy rotational requirements within the specified deploy-to-dwell ratios; alternatives with 30 PAA in AC and AFRC squadrons and 18 PAA in ANG squadrons were able to satisfy these requirements if the post-surge deploy-to-dwell ratios were applied during non-surge periods.

Figure 2.2. Ability of Alternative F-35 Beddowns to Satisfy Rotational Deployment Requirements

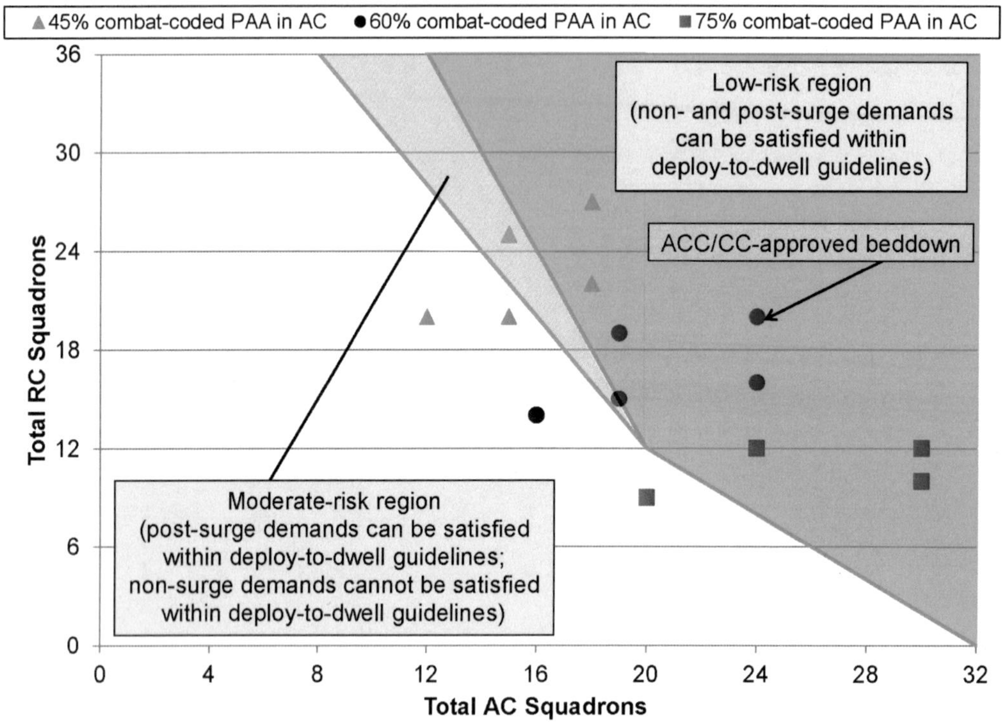

We recognize that this deployment demand analysis examines one view of the future environment, namely the ISCs, as provided by AF/A9. It is important to recognize that under an employment construct that differs from the one currently envisioned in the ISCs (in which the F-35 is basically deployed in a manner similar to the F-16), the deployment requirements and associated logistics resource requirements might differ significantly. Because the employment of the F-35 is still to be determined by the USAF, potential new concepts such as "many locations with very few F-35s at each location" could significantly change these requirements, and thus the supportability of an F-35 beddown that utilizes large squadron sizes.

3. Pilot Absorption

Background, Model Description, and "Feasible" Objectives for Absorption Analyses

Our analyses of pilot absorption capacities for the various beddown options were based on a steady-state absorption model that investigated potential "feasible" absorption conditions.[33] These conditions will: (1) provide enough pilots with adequate experience to generate sufficient pilot inventories, (2) use achievable aircraft utilization (UTE) rates,[34] and (3) maintain acceptable unit experience levels, while (4) enabling pilots to meet specified minimum Ready Aircrew Program (RAP) training requirements across all units in all components.

Model Description

Our absorption model calculated many of these required criteria based on previously agreed-upon input values. Thus we calculated the aircrew position indicator (API)-1 and API-6 pilot requirements in the combat-coded units—called *Force requirements* by the Air Force (Air Force Instruction [AFI] 11-401, AFI 38-201; and Thie, 1995)—because they changed with the beddown alternatives,[35] added in the *Non-Force* requirements (that do not change with beddown alternatives), and then checked them against inventories calculated in the models using the absorption results and historical data for bonus take rates (i.e., the rates at which AC pilots take the aviator continuation pay bonus and remain on active duty after their initial service commitment expires) and pilot affiliation rates with the RC for those who do not. Similarly, the model calculated the aircraft UTE required to ensure that all assigned pilots in all components could fly their minimum RAP requirements and used this information to calculate the aging rates and experience levels for all units.

---

[33] Absorption capacities measure the number of new pilots that operational units can absorb per year. Operational units absorb new fighter pilots by providing the training, experience, and supervision needed to develop them into combat pilots, instructors, and leaders. Important factors include unit manning and experience levels as well as facilities (e.g., simulators, ranges, and airspace) and aircraft utilization rates.

[34] For fighter aircraft, UTE is defined as the average number of sorties flown per PAA per month.

[35] API-1 positions are line pilots assigned to flying squadrons. API-6 positions are staff or supervisory billets assigned at wing-level or below that require incumbents to fly regularly.

The model input values can be broken down into three categories: (1) those dealing with pilot requirements, (2) those dealing with pilot inventories, and (3) those dealing with pilot training. They are summarized below: [36]

Pilot Requirements

- API-1s = smallest integer ≥ 1.25 X PAA per squadron
- AC API-6s = 2 combat mission ready (CMR) per squadron; 3 CMR per wing; 20 basic mission capable (BMC) per wing[37]
- RC API-6s = 2 CMR per squadron; 3 CMR per wing; 13 BMC per wing
- Non-Force requirements = 1990 for AC; 572 for RC

Pilot Inventories

- AC bonus take rate = 50 percent
- Bonus takers = 18 years of rated AC service
- Bonus nontakers = 10 years of rated AC service
- RC affiliation rate for bonus nontakers = 80 percent
- Affiliating pilots = 10 years RC rated service
- Pilots who serve only in the ANG or AFRC are assumed to have 18 years of RC rated service

Training Assumptions

- AC Fighter Training Unit output = 14 percent First Assignment Instructor Pilots and 86 percent Undergraduate Pilot Training (UPT) graduates[38]
- RAP minimums per month:[39]
  - Inexperienced CMR = 9/4 sorties/simulations for AC; 9/3 sorties/simulations for RC
  - Experienced CMR = 8/4 sorties/simulations for AC; 6/3 sorties/simulations for RC
  - BMC (all) = 5/2 sorties/simulations.

Other parameter values introduced in the discussion are model outputs. It should be noted that simulator training is an integral portion of pilot experiencing rates as well as

---

36 These assumptions were thoroughly discussed and vetted by the working group that advised this project. The training requirements for AC pilots were taken from a draft F-35 RAP Tasking Memorandum prepared by ACC/A3T, and those for RC pilots were adapted from the current F-16 Blk 50-52 RAP Tasking Memorandum, AS-12, effective 1 October 2011 provided via email.

37 CMR and BMC indicate a pilot's status for performing an aircraft's mission (for the F-16, for example, see Air Force Instruction [AFI] 11-2F-16 v1, 2011).

38 Most pilots who enter fighter training units are recent graduates of UPT; some graduates have a teaching position at UPT as their first assignment—they are First Assignment Instructor Pilots—and after their instructor assignment, some of those go on to fighter training. The absorption model has to make assumptions about how many fighter training unit students are recent UPT graduates and how many are instructors, as this affects how quickly they become experienced pilots.

39 RAP Tasking Memoranda for different aircraft provided via email describe the minimum monthly sortie requirements to maintain CMR or BMC status. These requirements vary depending on whether a pilot is experienced or inexperienced. Criteria for being "experienced" vary with the aircraft, but it generally means that the pilot has at least 500 hours flying the aircraft (AFI 11-2F-16 V1, 2011).

required RAP training, and the model assumes that pilots in all units are able to access high-fidelity concurrent simulators as required to obtain this training. The absorption computations would not be accurate without this simulator training.

Absorption Excursions

The historical norm for fighter pilot absorption has been to fill all AC inventory needs, plus all prior-service ANG and AFRC inventory needs, using pilots absorbed primarily in AC units. However, this has not been feasible since the late 1990s because the post–Cold War drawdown took AC force structure below the required levels (Taylor et al., 2002; Taylor et al., 2009). Indeed, not even the beddown alternative that put 75 percent of the combat-coded F-35s into the AC has sufficient absorption capacity to build adequate inventories using AC force structure alone. To increase absorption capacities for our analyses, staff from the Directorate of Operations, Deputy Chief of Staff for Operations, Plans and Requirements, Headquarters U.S. Air Force (AF/A3O), AFRC and NGB directed us to examine three distinct absorption excursions (for each of the 28 beddown options) using Active Associations, which consist of AC pilots operating ANG and AFRC airframes in ANG- and AFRC-assigned units. We were also asked to incorporate Classic Associations (with ANG and AFRC pilots flying with AC units) for every CONUS-based AC unit.

The principal difference between the excursions lay in their numbers and the mode by which AC pilots were assigned to ANG units (with embedded pilots included in the normal crew ratio authorizations and added pilots assigned over and above normal crew ratios):

Excursion 1:

- **AFRC squadrons**: two experienced and seven inexperienced AC CMR pilots, all embedded
- **ANG squadrons**: one experienced and three inexperienced AC CMR pilots embedded; five inexperienced AC CMR pilots added

Excursion 2:

- **AFRC squadrons**: two experienced and seven inexperienced AC CMR pilots, all embedded
- **ANG squadrons**: one experienced and three inexperienced AC CMR pilots embedded (no added pilots)

Excursion 3:

- **AFRC squadrons**: two experienced and seven inexperienced AC CMR pilots embedded per 24 PAA squadron, with embedded pilots increasing proportionally to PAA per squadron for options with PAA > 24

- **ANG squadrons**: one experienced and three inexperienced AC CMR pilots embedded per 18 PAA squadron, with embedded pilots increasing proportionally to PAA per squadron for options with PAA > 18

Sufficient Pilot Inventories Can Be Generated With RAP Minimum Overfly, But RC Experience and UTE Rate Would Not Maintain Traditional Levels

Only the first excursion above produced pilot inventories that approached the required levels, but even this excursion tended to impose a disproportionate share of the absorption burden on the ANG and AFRC units. Figures 3.1 and 3.2 present the resulting experience levels in ANG and AFRC units, respectively. Figures 3.3 and 3.4 present the resulting UTE requirements at RC and AC units, respectively. In these figures, the upper and lower limits of each bar correspond to the range of experience levels (or required UTE) achieved across all alternative beddowns corresponding to the specified combination of PAA per squadron and percent of total PAA in the AC.

RC experience levels increase as squadron size increases, and RC experience levels increase as the percentage of total PAA in the AC increases. Because the number of associated AC pilots per unit does not vary with RC squadron size in the first excursion, the inexperienced AC pilots have a lesser effect on the overall experience level for a larger RC squadron. As the percentage of aircraft in the AC increases, more new AC pilots are absorbed each year, which in turn generates a larger pool of AC pilots who eventually depart the AC as experienced pilots and affiliate with RC units, decreasing the RC units' requirement to train their own inexperienced non-prior-service pilots. Note that the experience levels presented in Figures 3.1 and 3.2 are significantly less than RC fighter units' historical experience levels, which have typically exceeded 85 percent. Indeed, ANG unit experience levels dropped below 60 percent for several beddown alternatives.

For RC units, we observed that the UTE requirement decreased as squadron size increases, but was not significantly affected by the percentage of total PAA in the AC. This is because the number of associated AC pilots does not vary with RC squadron size in the first excursion, and thus the increased flying needed to support the AC pilots is distributed over a larger number of aircraft in larger RC squadrons. The experience was different for AC units, for which we observed that the UTE requirement decreased as the percentage of total PAA in the AC increased, but was not significantly affected by squadron size. This is because the total AC pilot inventory requirement includes a large number of pilots who are not in F-35 operational units, but who are needed for other missions, such as test and training squadrons, or staff positions. This Non-Force requirement for AC pilots outside of the F-35 operational units was assumed to be constant across all beddown alternatives; thus, alternatives with less aircraft in the AC have fewer AC units through which to absorb the total pilot requirement, whereas alternatives with more aircraft in the AC have a broader base of AC units through which the nonoperational units' fighter pilot requirements can be absorbed. In particular, we observed that ANG units have a required UTE rate that was two to three sorties per PAA per month (15 to 23 percent) greater than the AC UTE for many beddown alternatives.

Although the additional five inexperienced AC pilots imposed a significant burden on the ANG units in terms of UTE rates and experience levels, the other absorption excursions could not provide adequate pilot inventories. Within the first absorption excursion, six of the 28 beddown alternatives, each of which assumed 45 percent of combat-coded PAA in the AC and 24 PAA per ANG squadron, did not generate the required pilot inventories, with shortfalls ranging from 31 to 98 pilots. These pilot shortfalls can be eliminated if AC units are allowed to overfly the RAP minimums (the corresponding AC overfly for these beddown alternatives ranged between 2.4 and 7.5 percent above the RAP minimums). Thus, decisions to limit AC unit flying hours to levels that will ensure pilots cannot exceed RAP minimums[40] impose significant constraints on the units that historically have provided the principal engine for absorbing and developing new fighter pilots.[41]

Figure 3.1. ANG Experience Level, Under First Absorption Excursion

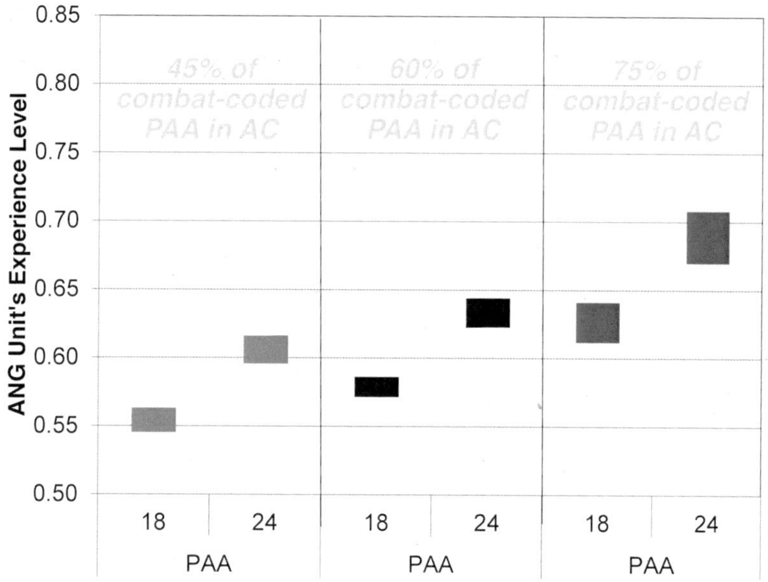

---

[40] Personal communication from Dean Gould, ACC/A3TB, October 9, 2012

[41] A discussion of the impact of sortie availability and squadron size on the ability to produce experienced pilots can be found on pages 59–62 of Taylor et al., 2002.

Figure 3.2. AFRC Experience Level, Under First Absorption Excursion

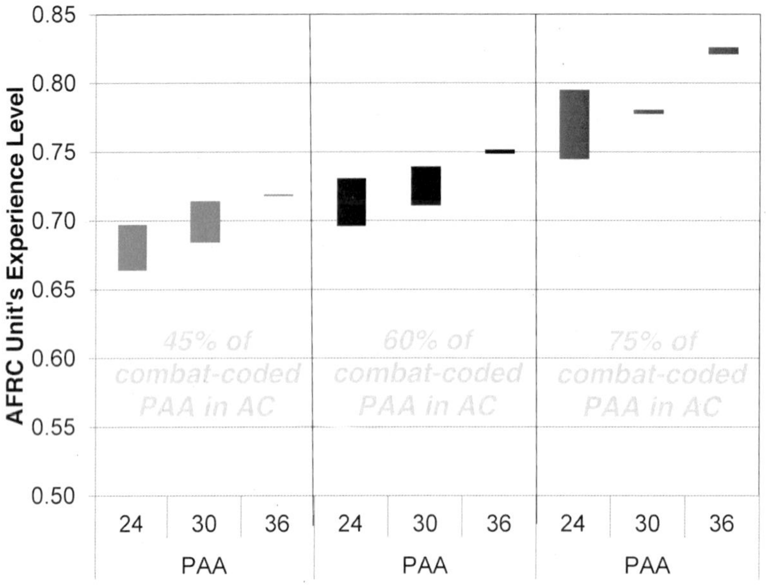

Figure 3.3. RC UTE Requirements, Under First Absorption Excursion

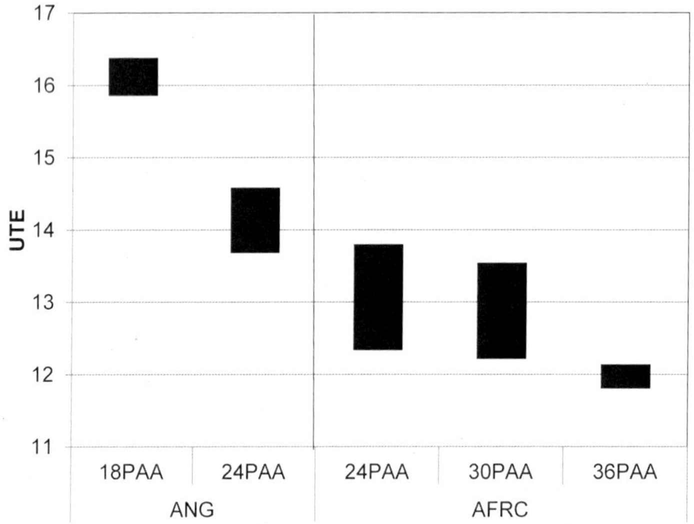

Figure 3.4. AC UTE Requirements, Under First Absorption Excursion

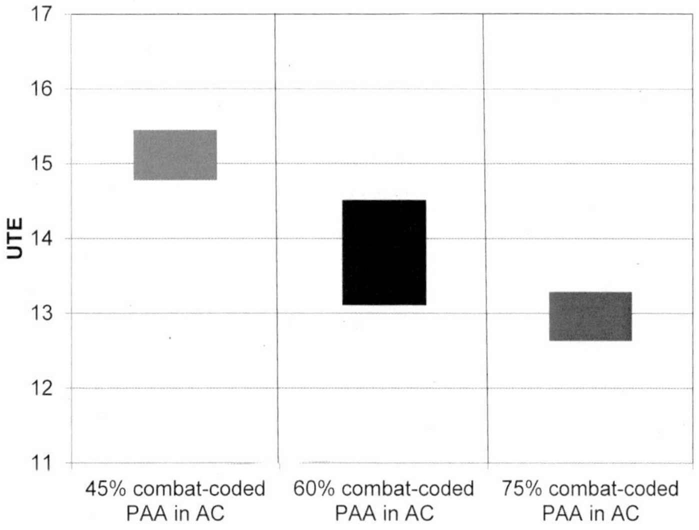

Traditionally AC F-16 units have flown at higher UTE rates than the corresponding Guard or Reserve units, and we have already discussed the traditional role of the AC units as the principal engine for absorbing and developing new fighter pilots—not only for the AC, but also for prior-service fighter pilots historically relied upon by the ANG and AFRC. AC units traditionally have larger squadrons and more squadrons per wing than RC units, which enable them to be more "absorption efficient" than the RC units. RC units typically have one squadron per wing and thus have to support more pilots per airframe than AC units. It would be extremely difficult to preserve these traditions for the F-35 while implementing a beddown plan that puts only 45 percent of the primary mission aircraft in the AC and places the majority of the force structure in the ANG and AFRC.

The absorption advantages traditionally enjoyed by AC units, however, cease to exist when units' flying hours are constrained to prevent pilots from flying more than their RAP minimums. These training requirements are specified as minimum requirements, after all, and it requires perfect scheduling and sortie management to ensure they are met precisely by all unit pilots. Resources to overfly RAP minimums will be required in order for combat air forces (CAF) units to "fly their way out" of pilot absorption problems.[42]

---

[42] For a brief discussion of how the Air Force addressed pilot absorption shortfalls arising from the post-Vietnam-era drawdown, see Taylor et al., 2009. For a more comprehensive view, see Anderegg, 2001.

Impact on Pilot Absorption Flying Costs

Under the first absorption excursion (given the set of UTE requirements identified for each of the 28 alternative beddowns), we identified the annual cost associated with generating the required number of sorties, assuming an average sortie duration of 1.4 flying hours and a cost of $18,025 per flying hour.[43] Figure 3.5 illustrates the annual pilot absorption costs associated with each of the 28 beddown alternatives, with the value for each alternative presented as the percentage difference between its cost and the cost of the baseline ACC/CC-approved beddown. Each marker on the figure corresponds to one paired set of beddown alternatives—each member of the set has an equal number of RC and AC squadrons, they differ only in the percentage of total AC PAA in CONUS (which did not have a significant impact on the costs presented here).

As the fraction of combat-coded PAA in the AC is held constant (i.e., within the set of circles, squares or triangles in the figure), increasing the squadron size (i.e., moving down and to the left on the figure, with fewer squadrons) can significantly reduce the annual pilot absorption flying cost. Observe that the ACC/CC-approved beddown has an annual pilot absorption flying cost of $4.4 billion. Within the alternative that maintains 60 percent of the combat-coded PAA in the AC, increasing AC and AFRC squadron size to 30 PAA while maintaining 18 PAA per ANG squadron (represented in the chart by the black circle at the point with 19 AC squadrons and 19 RC squadrons) can reduce these costs 4 percent relative to the ACC/CC-approved beddown, while increasing ANG squadron size to 24 PAA and maintaining 24 PAA per AC and AFRC squadron (represented in the chart by the black circle at the point with 24 AC squadrons and 16 RC squadrons) could reduce these costs by 8 percent. Increasing both AC and AFRC squadrons to 30 PAA and ANG squadrons to 24 PAA (represented in the chart by the black circle at the point with 19 AC squadrons and 15 RC squadrons) would reduce these costs by 10 percent, while a further increase to 36 PAA in the AC and AFRC could reduce these costs by 12 percent.

---

[43] Air Force Cost Analysis Agency (AFCAA) provided us with an F-35A Steady State cost per flying hour (CPFH) in Base Year 2012 dollars. "Steady state" is defined here as the average cost during the period with the maximum number of PAA, which for the F-35A is FY2036–2040. This factor includes cost growth above inflation, and is composed of costs for fuel ($6,604), consumables ($1,793) and depot-level repairables ($9,628).

Figure 3.5. Annual Pilot Absorption Flying Costs, by Beddown Alternative

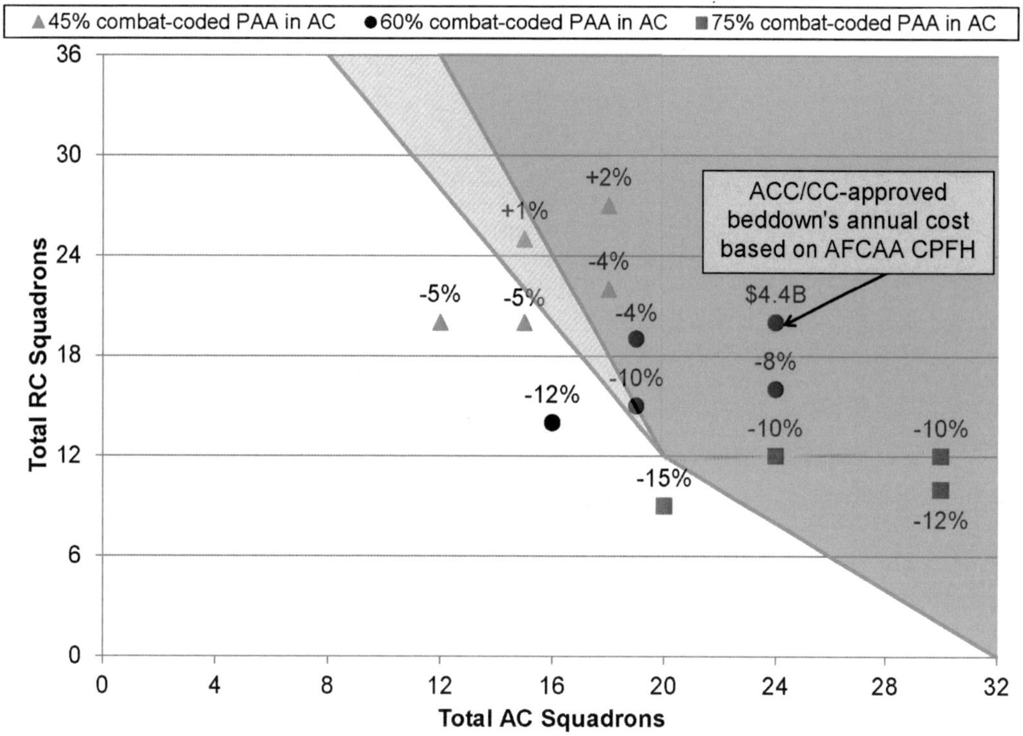

NOTE: The dark green region on this figure corresponds to the range over which all non-surge and post-surge demands can be satisfied within the deploy-to-dwell ratios identified in Table 2.2. The light green region on this figure corresponds to the range over which all rotational requirements can be satisfied within the post-surge deploy-to-dwell ratios.

As the squadron size is held constant, increasing the fraction of combat-coded PAA in the AC (e.g., comparing the marker farthest to the left for each colored set of markers) also generates cost reductions. When compared to the ACC/CC-approved beddown's $4.4 billion in annual pilot absorption flying costs, there are many alternative beddowns that satisfy all deployment requirements and reduce this cost by 10 percent or more.

Finding Feasible Results

Because none of the three absorption excursions that we were asked to analyze generated a feasible absorption condition that also maintained the RC's higher experience levels and lower UTE rates, relative to the AC, we performed additional analysis to search for a set of assumptions that would satisfy all of these objectives. These additional provisions lead to a relatively simple feasible absorption condition:

- **AFRC squadrons**: Two experienced and seven inexperienced AC CMR pilots embedded per 24 PAA squadron, with embedded pilots increasing proportionally to PAA per squadron for options with PAA > 24 (identical to excursion 3 above).
- **ANG squadrons**: Two experienced and seven inexperienced AC CMR pilots embedded per 24 PAA squadron, with embedded pilots decreasing proportionally to PAA per squadron for options with PAA = 18 (similar to AFRC).

If a 10 percent overfly of RAP minimums is enforced at all AC units, all beddown alternatives that have at least 60 percent of the F-35 combat-coded PAA in the AC remain feasible, regardless of whether ANG units contain one or two experienced AC supervisors. Beddown alternatives with 45 percent of the combat-coded PAA in the AC require an overfly of RAP minimums by 10 to 16 percent at AC units to generate sufficient pilot inventories. AC UTE rates compare favorably with the UTE rates required in ANG and AFRC units in the sense that in all but one case they are at least as large or larger. The required UTE rates range from 13.4 to 14.8 for AC units and from 12.2 to 14.0 in RC units.[44] If decisionmakers desire to maintain a more traditional UTE relationship among the components' units, with a lower UTE in RC units than in AC units, they will need to accept a higher allowable overfly of RAP minimums at AC units.

In many cases, the total annual flying hours required under this feasible excursion compare favorably with the requirements of the first absorption excursion. For corresponding beddowns with 45 percent of combat-coded PAA in the AC, the feasible excursion's pilot absorption flying cost is between 5.8 and 8.3 percent less than the first excursion. For corresponding beddowns with 60 percent of combat-coded PAA in the AC, the cost is between 0.4 and 3.6 percent less, and for corresponding beddowns with 75 percent of combat-coded PAA in the AC, the cost is between 0.6 and 2.8 percent greater.

This is a single feasible result that we readily identified once the problem's bounds were adequately relaxed. Additional analysis would be useful to better understand the set of feasible solutions and to pursue feasible results that are optimal in meaningful contexts for Air Force leaders in all components.

We note that these absorption results are not meaningful unless adequate simulators are available to the pilots in all units. It is imperative that the final beddown plan includes readily accessible high-fidelity concurrent simulator facilities to provide essential pilot experience and RAP training opportunities.

[44] Again, UTE rates are measured as sorties per PAA per month.

4. Logistics Resources

F-35 Maintenance Manpower Requirements

The first logistics resource that we examined was maintenance manpower. As of February 2012, F-35 squadron maintenance manpower requirements, as described in the JPO Manpower Estimate, existed only for a squadron size of 24 PAA.[45] ACC provided us with the detailed maintenance manpower requirements for a combat-coded squadron of 24 PAA.[46] These data classify each manpower position as either a *direct* or an *indirect* position. *Direct* maintenance spaces are those for which the manufacturer can influence the requirement based on the aircraft's expected reliability and maintainability. Direct maintenance includes most "touch labor" work centers, such as the Aircraft Section (Crew Chiefs) or Egress maintenance. The requirement for *indirect* maintenance spaces is based on USAF policy and organizational structure. Indirect maintenance includes most supervisory and support work centers, such as Maintenance Analysis or Maintenance Training. As of December 2011, ACC did not have any information regarding maintenance manpower requirements for F-35 squadrons of other PAA values.

However, these data do not reflect the entirety of maintenance manpower requirements at a squadron. There are a large number of additional positions in supervisory and support roles that are needed to support a squadron of aircraft but are not included in the list of F-35 maintenance positions provided by ACC.

We utilized ACC and Air Force Manpower Agency (AFMA) guidance documents to identify these additional supervision and support position requirements,[47] and thus generate a total maintenance manpower requirement for an F-35 combat-coded squadron of 24 PAA. In some instances, we added positions to work centers in the Maintenance Group and Maintenance Operations Squadron above the values provided by ACC in order to be consistent with the guidance documents. In such cases, we identify these additional manpower positions as a separate work center, indicated with a leading asterisk (*) in the work center name. These total requirements are detailed in Table 4.1.

---

[45] We note that there is a discrepancy between this assumption and the 18 PAA ANG squadrons in the ACC/CC plan.

[46] Email from Howard Beizer ACC/A1MPP, received October 7, 2011.

[47] We utilized the spreadsheet "F-16B40 Estimator.xlsx", provided by the Second Manpower Requirements Squadron, to generate these manpower requirements. This spreadsheet states that it utilizes source reference documents from ACC (2011), and the U.S. Air Force (1997a, 1997b, 2008, 2002).

Note that while there are only 267 "direct" maintenance positions required in an F-35 squadron of 24 PAA, such a squadron has a total requirement of 547 maintenance manpower positions, with supervisory and support positions and munitions maintenance accounting for most of the difference.

Table 4.1. Total Maintenance Manpower Requirements for an F-35 Combat-Coded Squadron of 24 PAA

Work Center	Manpower	Direct Maintenance Position?	Not Included in ACC-Provided Requirement?
Maintenance Group			
Command Section	7		X
Weapons Standardization	2		X
Lead Weapons Crew	3	X	
Quality Assurance	12		
* Additional Quality Assurance	3		X
Maintenance Operations Squadron			
Command Section	3		X
Maintenance Operations Flight Supervision	1		X
Maintenance Analysis	2		
* Additional Maintenance Analysis	6		X
Engine Management	4		X
Maintenance Operations Center	4		
* Additional Maintenance Operations Center	10		X
Plans, Scheduling & Documentation	1		
* Additional Plans, Scheduling & Documentation	4		X
Programs and Resources	4		X
Maintenance Training	4		
* Additional Maintenance Training	4		X
Aircraft Maintenance Squadron			
Maintenance Supervision	9		X
Command Section	3		X
Aircraft Maintenance Unit	19		

Work Center	Manpower	Direct Maintenance Position?	Not Included in ACC-Provided Requirement?
Aircraft Section Supervision	4		
Aircraft Section (Crew Chiefs)	70	X	
Alert-Crew Chiefs	15		
End of Runway-Crew Chiefs	6		
Specialist Supervision[a]	2		
Avionics Specialists	42	X	
Weapons Supervision[b]	2		
Weapons Loading	45	X	
Weapons Maintenance	15	X	
End of Runway-Weapons	6		
Support Section	18		
Maintenance Squadron			
Maintenance Supervision	7		X
Command Section	3		X
Accessories Flight Supervision	1		X
Fuel Systems Supervision	1		X
Fuel Systems	12	X	
Egress Supervision	1		X
Egress	8	X	
Fabrication Flight Supervision	1		X
Non-Destructive Inspection (NDI) Supervision	1		X
NDI	11	X	
Aircraft Structural Maintenance Supervision	1		X
Aircraft Structural Maintenance	40	X	
Maintenance Flight Supervision	1		X
Aircraft Inspection Supervision	1		X
Aircraft Inspection	15	X	
Wheel and Tire Supervision	2		X
Wheel and Tire	6	X	
Aerospace Ground Equipment[c]	25		
Munitions	80		
Totals			
Direct maintenance positions		**267**	
Not included in ACC-provided requirement		**80**	
All positions		**547**	

[a] We used the value presented by ACC as the F-35 requirement, but note that this value is inconsistent with *ACC Aircraft Maintenance Command Guide* (2011), Paragraph 5.3.2.2, which states a requirement of four positions.

[b] We used the value presented by ACC as the F-35 requirement, but note that this value is inconsistent with *ACC Aircraft Maintenance Command Guide* (2011), Paragraph 5.3.2.3, which states a requirement of four positions.

[c] We used the value presented by ACC as the F-35 requirement, but note that this value is inconsistent with Air Force Manpower Standard 23F1 (1997b), which states a requirement of 32 positions for the F-16.

Maintenance Manpower Requirements for F-35 Squadrons of Varying PAA

For legacy USAF aircraft, such as the F-16, one could determine the maintenance manpower requirement for a squadron of any PAA level by utilizing the Logistics Composite Model (LCOM), a stochastic simulation model most commonly used to determine the manpower requirements associated with aircraft maintenance activities.[48] These activities include the preparation of aircraft on the flightline, the repair of planes and aircraft components that experience a failure during flight operations, and the scheduled maintenance of airframes.

However, the USAF does not currently have an LCOM for the F-35, because it has not yet been possible to perform the extensive data collection and audits necessary to determine the failure rate and repair time data required to run the model.[49] We instead utilized F-16 and F-22 LCOM runs to provide insight into the relationship between maintenance manpower requirements and squadron PAA.

The 2nd Manpower Requirements Squadron provided us with LCOM simulation results for both the F-16 and F-22 for a range of squadron PAA values. Table 4.2 presents these data, which generally do not include supervisory and support positions, but instead present the "touch labor" maintenance requirement for each work center.[50] We summed the Crew Chief and Flightline Propulsion Specialist manpower values for each weapon system-PAA pair, because the F-35 flightline engine maintenance workload will be performed by F-35 crew chiefs. Similarly, the Avionics Specialist and Electronics and Environmental (E/E) Specialist manpower values were summed for each weapon system-PAA pair, because all of these workloads will be performed by F-35 flightline avionics specialists.

---

[48] LCOM was initially developed by RAND (Fisher, 1968).

[49] Dynamics Research Corporation built an F-35 LCOM for the JPO in 2004, using 2004 reliability and maintainability estimates. However, these data have not been updated since that time and are viewed by ACC/A1/A4 staffs as not useful for generating new F-35 maintenance manpower requirements, in light of changes to the F-35 program that have occurred during the intervening years.

[50] The exceptions are Aerospace Ground Equipment (AGE) and Munitions Maintenance, for which the Table 4.2 manpower totals include both maintenance and supervisory and support positions, because the F-35 data that we received for these work centers in Table 4.1 present a total manpower requirement and do not separately identify the supervisory and support requirement.

Table 4.2. LCOM-Based Maintenance Manpower Requirements for F-16 and F-22 Combat-Coded Squadrons of Varying PAA

Weapon System	F-16				F-22			
PAA	18	24	30	36	18	24	30	36
Crew Chiefs + Flightline Propulsion Specialists[a]	107	135	164	192	68	87	97	106
Avionics Specialists + E/E Specialists	40	50	59	68	37	46	49	52
Weapons Loading	42	54	63	73	45	54	59	63
Weapons Maintenance[b]	15	18	22	26	-	-	-	-
Fuel Systems	15	18	22	26	11	17	19	21
Egress	12	15	19	23	9	12	14	15
NDI	8	9	11	13	12	15	15	15
Aircraft Structural Maintenance	21	25	29	33	56	74	83	93
Aircraft Inspection	23	27	32	36	0	0	0	0
Wheel and Tire	7	8	9	10	3	3	3	3
Aerospace Ground Equipment	34	40	45	52	35	35	35	35
All Munitions Maintenance, except for Armament	90	97	106	116	113	134	152	163

[a] For the remainder of this analysis, we present the total of all nonsupervisory crew chief positions in the Aircraft Maintenance Squadron [Aircraft Section (Crew Chiefs), Alert-Crew Chiefs and End of Runway-Crew Chiefs]. Note that the F-22 data we received did not identify the nonsimulated requirements for Alert-Crew Chiefs and End of Runway-Crew Chiefs; thus the F-22 data presented here are the LCOM-simulated Aircraft Section (Crew Chiefs) manpower values only.
[b] For the remainder of this analysis, we present the total of all nonsupervisory weapons maintenance positions in the Aircraft Maintenance Squadron (Weapons Maintenance and End of Runway-Weapons). Note that the F-22 data we received did not differentiate between Weapons Loading and Weapons Maintenance work centers, but instead presented a total Weapons manpower requirement. Moreover, these F-22 data did not identify the nonsimulated requirements for End of Runway-Weapons; thus the F-22 data presented here are the LCOM-simulated Weapons manpower values only.

Prior analyses of these (and other) weapon systems have demonstrated that maintenance manpower requirements exhibit an *economy of scale* with respect to squadron PAA: squadrons with more PAA typically require less maintenance manpower, on a per-PAA basis, than do squadrons with fewer PAA. Table 4.3 presents the maintenance manpower requirements per PAA for this set of F-16 and F-22 work centers. We observe the existence of economies of scale in each work center for each weapon system, since the manpower required per PAA decreases as squadron size increases.[51]

[51] Only two cells in Table 4.3 are inconsistent with this general trend: F-16 Egress at 36 PAA, whose 0.64 maintenance manpower spaces per PAA is slightly larger than the corresponding value of 0.63 at 30 PAA,

Table 4.3. Maintenance Manpower Requirements Per PAA for F-16 and F-22 Combat-Coded Squadrons of Varying PAA

Weapon System	**F-16**				**F-22**			
PAA	**18**	**24**	**30**	**36**	**18**	**24**	**30**	**36**
Crew Chiefs + Flightline Propulsion Specialists	5.94	5.63	5.47	5.33	3.78	3.63	3.23	2.94
Avionics Specialists + E/E Specialists	2.22	2.08	1.97	1.89	2.06	1.92	1.63	1.44
Weapons Loading	2.33	2.25	2.10	2.03	2.50	2.25	1.97	1.75
Weapons Maintenance	0.83	0.75	0.73	0.72	-	-	-	-
Fuel Systems	0.83	0.75	0.73	0.72	0.61	0.71	0.63	0.58
Egress	0.67	0.63	0.63	0.64	0.50	0.50	0.47	0.42
NDI	0.44	0.38	0.37	0.36	0.67	0.63	0.50	0.42
Aircraft Structural Maintenance	1.17	1.04	0.97	0.92	3.11	3.08	2.77	2.58
Aircraft Inspection	1.28	1.13	1.07	1.00	-	-	-	-
Wheel and Tire	0.39	0.33	0.30	0.28	0.17	0.13	0.10	0.08
Aerospace Ground Equipment	1.89	1.67	1.50	1.44	1.94	1.46	1.17	0.97
All Munitions Maintenance, except for Armament	5.00	4.04	3.53	3.22	6.28	5.58	5.07	4.53

There are two primary reasons for the larger squadrons' increased efficiency.[52] First, for the smaller squadron sizes considered here (18 to 24 PAA), many work centers are influenced by *minimum crew size* effects, where their manpower requirement is determined by the work center task that requires the largest crew size, even if most work center tasks require a smaller crew. Second, all work centers need to be sized to account for "insurance" effects, since the maintenance organization needs the capacity to accommodate random spikes in demand without a significant disruption to flying operations. At larger squadron sizes, the impact of variations in both demand and the duration of maintenance activities is dampened by the aggregation of demands across a larger number of aircraft, allowing work centers to increase the maximum manpower utilization that can be supported without generating significant queues of unavailable components and aircraft.

and F-22 Fuel Systems at 18 PAA, whose 0.61 maintenance manpower spaces per PAA is less than the corresponding values of 0.71 and 0.63 at 24 PAA and 30 PAA, respectively.

[52] For the AGE and Munitions data in Table 4.3, an additional factor is that larger squadrons are able to spread the relatively constant supervisory and support requirements, which are generally not very sensitive to the number of PAA in the squadron, across a larger number of aircraft.

We can use these same data to identify a *manpower growth rate* for each weapon system/work center/PAA value, defined as the corresponding manpower requirement divided by the manpower requirement for that weapon system/work center at 24 PAA. As an example, the manpower growth rate for F-16 Aircraft Inspection at 36 PAA is computed as *36 / 27 = 1.33.* Table 4.4 presents the manpower growth rates for this set of F-16 and F-22 work centers.

Table 4.4. Work Center Manpower Growth Rate, Versus 24 PAA Baseline, for F-16 and F-22 Combat-Coded Squadrons of Varying PAA

Weapon System	F-16			F-22		
PAA	**18**	**30**	**36**	**18**	**30**	**36**
Crew Chiefs + Flightline Propulsion Specialists	0.79	1.21	1.42	0.78	1.11	1.22
Avionics Specialists + E/E Specialists	0.80	1.18	1.36	0.80	1.07	1.13
Weapons Loading	0.78	1.17	1.35	0.83	1.09	1.17
Weapons Maintenance	0.83	1.22	1.44	-	-	-
Fuel Systems	0.83	1.22	1.44	0.65	1.12	1.24
Egress	0.80	1.27	1.53	0.75	1.17	1.25
NDI	0.89	1.22	1.44	0.80	1.00	1.00
Aircraft Structural Maintenance	0.84	1.16	1.32	0.76	1.12	1.26
Aircraft Inspection	0.85	1.19	1.33	-	-	-
Wheel and Tire	0.88	1.13	1.25	1.00	1.00	1.00
Aerospace Ground Equipment	0.85	1.13	1.30	1.00	1.00	1.00
All Munitions Maintenance, except for Armament	0.93	1.09	1.20	0.84	1.13	1.22

Observe that the manpower growth rate is in most cases larger for the F-16 than for the F-22. For six work centers (Crew Chiefs + Flightline Propulsion Specialists, Avionics Specialists + E/E Specialists, Fuel Systems, Egress, NDI, and Aircraft Structural Maintenance) the F-16 manpower growth rate is greater than the F-22 rate for all PAA values considered. For three work centers (Weapons Loading, Wheel and Tire, and Aerospace Ground Equipment) the F-16 manpower growth rate is greater than the F-22 value for 30 PAA and 36 PAA squadrons, but less than the F-22 rate for 18 PAA squadrons. There is only one work center (All Munitions Maintenance, except for Armament) for which the F-16 manpower growth rate is less than the F-22 rate for 30 PAA or 36 PAA squadrons.

In order to estimate the maintenance manpower requirements at F-35 squadrons of different PAA values, we will assume that the relationship observed between squadron size and maintenance manpower requirements for legacy fighter aircraft will apply to the F-35. For each "touch labor" work center, we apply the F-16 growth rates presented in Table 4.4 to the F-35 manpower requirements presented in Table 4.1 to compute an F-35 requirement. For example, the ACC-provided data in Table 4.1 state that an F-35 squadron of 24 PAA requires 11 NDI positions. To estimate the requirement for an F-35 squadron of 36 PAA, we multiply the F-16 manpower growth rate associated with NDI-

36 PAA (1.44) times the F-35 requirement at 24 PAA (11) and, rounding up to the next integer value, obtain a requirement of 16 NDI positions.

We use the F-16 manpower growth rates here because the F-22 has, in most instances, smaller manpower growth rates and would thus generate a smaller F-35 manpower requirement. Had we instead used the F-22 rates, our estimated F-35 manpower requirements would have been reduced by seven positions for 18 PAA, 25 positions for 30 PAA, and 57 positions for 36 PAA. We utilized the F-16 rates in order to make conservative estimates regarding the economies of scale that could be achieved in F-35 maintenance manpower were the PAA per squadron increased.

For supervisory and support positions, we utilized the ACC and AFMA guidance documents discussed previously to identify manpower requirements for alternative squadron sizes. In many cases, these documents provided sufficient guidance to identify a requirement as a function of squadron size. For example, Fabrication Flight Supervision manpower is based upon the manpower in subordinate work centers (for the F-35, these work centers are NDI and Aircraft Structural Maintenance): If there are between 18 and 67 positions in subordinate work centers, one position is earned; if there are between 68 and 126 positions, two positions are earned, and so on (ACC, 2011).

However, there were some work centers, such as the Aircraft Maintenance Unit (AMU), for which it is unclear whether the guidance documents are applicable to the largest squadron sizes under consideration. The guidance document for AMU manpower requirements identifies a requirement for 19 AMU positions per Aircraft Maintenance Squadron (AMXS), independent of the number of PAA supported by the squadron (ACC, 2011). At the time that this guidance document was written, no ACC fighter squadrons had more than 24 PAA; thus, it is unclear whether this AMU requirement would apply to squadron sizes that are larger than 24 PAA.

In order to address instances in which the guidance documents might not provide an accurate count of supervisory and support position requirements for squadrons of more than 24 PAA, we shared our initial estimates for these positions, based on the exact language appearing in the guidance documents, with subject matter experts at the Directorates of Logistics at both Headquarters, U.S. Air Force (AF/A4L) and Headquarters, Air Combat Command (ACC/A4). Staff members at both organizations reviewed these estimates and suggested which supervisory and support work centers would need to have their manpower requirements increased in order to support 36 PAA, the largest squadron size under consideration. Based upon their reviews, we added 18 supervisory and support positions per AMXS for 36 PAA.[53] For squadrons of 30 PAA, we added one-half of this increase to each affected work center.[54]

The total manpower requirements that we estimated for each F-35 squadron size appear in Table 4.5.

Table 4.5. Total Maintenance Manpower Requirements for F-35 Combat-Coded Squadrons of Varying PAA

Work Center	Manpower at 18 PAA	Manpower at 24 PAA	Manpower at 30 PAA	Manpower at 36 PAA
Maintenance Group				
Command Section	7	7	7	7
Weapons Standardization	5	5	5	5
Quality Assurance	15	15	15	15
Maintenance Operations Squadron				
Command Section	3	3	3	3
Maintenance Operations Flight Supervision	1	1	1	1
Maintenance Analysis	7	8	8	9
Engine Management	4	4	4	4
Maintenance Operations Center	14	14	14	18
Plans, Scheduling & Documentation	4	5	5	5
Programs and Resources	4	4	4	4
Maintenance Training	8	8	9	10
Aircraft Maintenance Squadron				
Maintenance Supervision	8	9	10	10
Command Section	3	3	3	3
Aircraft Maintenance Unit	19	19	23	27
Aircraft Section Supervision	4	4	5	6
Aircraft Section (Crew Chiefs)	72	91	111	129
Specialist Supervision	2	2	2	2
Avionics Specialists	34	42	50	57
Weapons Supervision	2	2	3	4
Weapons Loading	35	45	53	61
Weapons Maintenance	18	21	26	30
Support Section	17	18	20	26
Maintenance Squadron				
Maintenance Supervision	7	7	8	8
Command Section	3	3	3	3

53 These positions were composed of eight additional AMU positions, two additional Aircraft Section Supervision positions, two additional Weapons Supervision positions, and six additional Support Section positions.

54 The Support Section was an exception. Unlike the other three work centers for which manpower was increased based upon AF/A4L and ACC/A4 review, the Support Section requirement identified in the guidance documents increased from 18 positions in support of 24 PAA to 20 positions in support of 30 PAA or 36 PAA. We assumed that this growth from 18 to 20 positions was sufficient to support the increased workload at 30 PAA.

Work Center	Manpower at 18 PAA	Manpower at 24 PAA	Manpower at 30 PAA	Manpower at 36 PAA
Accessories Flight Supervision	1	1	1	1
Fuel Systems Supervision	1	1	1	1
Fuel Systems	10	12	15	17
Egress Supervision	1	1	1	1
Egress	6	8	10	12
Fabrication Flight Supervision	1	1	1	2
NDI Supervision	1	1	1	1
NDI	10	11	13	16
Aircraft Structural Maintenance Supervision	1	1	1	1
Aircraft Structural Maintenance	34	40	46	53
Maintenance Flight Supervision	1	1	1	1
Aircraft Inspection Supervision	1	1	1	1
Aircraft Inspection	13	15	18	20
Wheel and Tire Supervision	2	2	2	2
Wheel and Tire	5	6	7	8
Aerospace Ground Equipment	21	25	28	33
Munitions	74	80	87	96
Totals	**479**	**547**	**626**	**713**

To estimate the maintenance manpower requirement at multisquadron wings, we utilized the ACC and AFMA guidance documents to identify the additional manpower requirement in each Maintenance Group (MXG) and Maintenance Operations Squadron (MOS) work center, and for the Maintenance Supervision and Command Section work centers in each of the AMXS and Maintenance Squadron (MXS). We assumed that all other AMXS and MXS work centers would have their manpower (as presented in Table 4.5) replicated for each squadron in the wing. Table 4.6 presents the total maintenance manpower that we estimated for each F-35 wing under consideration in this analysis.

Figure 4.1 presents these same manpower counts, now normalized on the basis of maintenance manpower requirements per PAA.[55] The figure shows that, for combat-coded aircraft, the required maintenance manpower per PAA decreases as the number of PAA per squadron increases. A single squadron of 18 PAA would require 26.6 maintenance positions per PAA, where a squadron of 36 PAA could be supported by 19.8 maintenance positions per PAA—26 percent fewer. Assigning multiple squadrons to a single wing can generate additional efficiencies beyond those generated by the squadron size effect. While a single squadron of 36 PAA requires 19.8 maintenance positions per PAA, a wing of three squadrons of 36 PAA each requires only 18.6 maintenance

[55] Note that the single option with four squadrons per wing is not graphed in Figure 2.1.

positions per PAA—a 6 percent reduction. Taken together, these effects suggest that increasing the F-35 squadron size and consolidating squadrons into multisquadron wings could generate significant reductions in maintenance manpower requirements.

Table 4.6. Total Maintenance Manpower Requirements for F-35 Combat-Coded Wings

Number of Squadrons per Wing	PAA per Squadron			
	18	24	30	36
1	479	547	626	713
2	n.a.	1,033	1,190	1,357
3	n.a.	1,514	1,749	2,004
4	n.a.	1,993	n.a.	n.a.

Figure 4.1. F-35 Maintenance Manpower Requirements per PAA

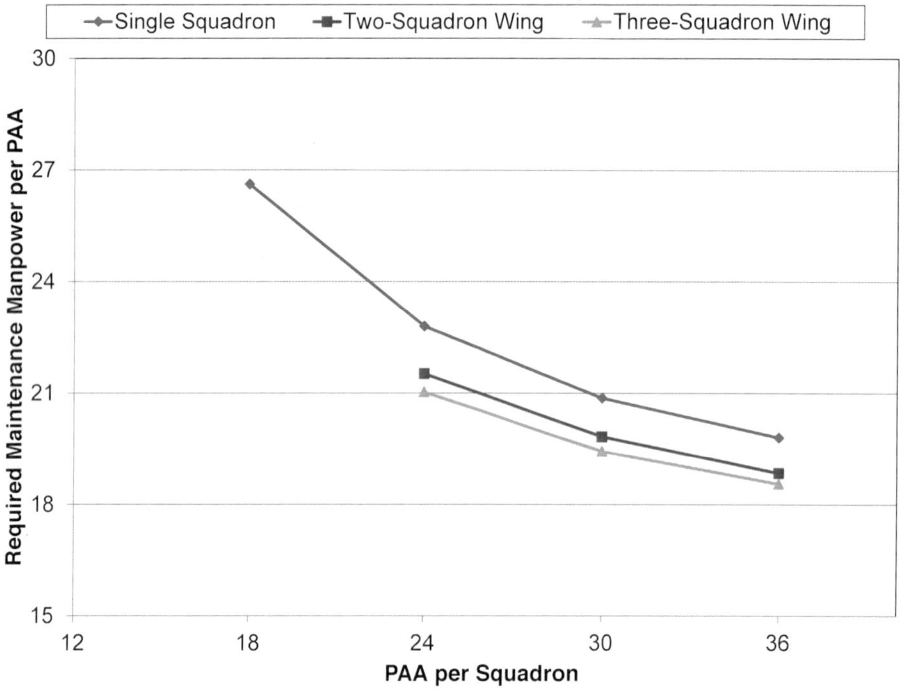

To demonstrate the source of efficiencies associated with the squadron size effect, consider two alternatives for assigning 72 PAA to a single wing: either in three squadrons of 24 PAA, or in two squadrons of 36 PAA. The wing that is arranged into two squadrons of 36 PAA requires 157 fewer positions, a decrease of 10 percent with respect to the 1,514 positions required in a wing with three squadrons of 24 PAA. Table 4.7 presents the differences in manpower, by work station, for these two alternatives.

Table 4.7. Total Maintenance Manpower Requirements for Two Alternative Combat-Coded F-35 Wing Structures, Each with 72 PAA

Work Center	Manpower with Three 24 PAA Squadrons	Manpower with Two 36 PAA Squadrons	Manpower Savings with 36 PAA Squadrons
Maintenance Group			
Command Section	8	7	1
Weapons Standardization	5	5	0
Quality Assurance	45	30	15
Maintenance Operations Squadron			
Command Section	3	3	0
Maintenance Operations Flight Supervision	1	1	0
Maintenance Analysis	12	12	0
Engine Management	4	4	0
Maintenance Operations Center	22	22	0
Plans, Scheduling & Documentation	6	6	0
Programs and Resources	4	4	0
Maintenance Training	17	15	2
Aircraft Maintenance Squadron			
Maintenance Supervision	14	14	0
Command Section	3	3	0
Aircraft Maintenance Unit	57	54	3
Aircraft Section Supervision	12	12	0
Aircraft Section (Crew Chiefs)	273	258	15
Specialist Supervision	6	4	2
Avionics Specialists	126	114	12
Weapons Supervision[a]	6	8	-2
Weapons Loading	135	122	13
Weapons Maintenance	63	60	3
Support Section	54	52	2
Maintenance Squadron			
Maintenance Supervision	14	12	2
Command Section	3	3	0
Accessories Flight Supervision	3	2	1
Fuel Systems Supervision	3	2	1
Fuel Systems	36	34	2
Egress Supervision	3	2	1
Egress	24	24	0
Fabrication Flight Supervision	3	4	-1
NDI Supervision	3	2	1
NDI	33	32	1
Aircraft Structural Maintenance Supervision	3	2	1
Aircraft Structural Maintenance	120	106	14
Maintenance Flight Supervision	3	2	1
Aircraft Inspection Supervision	3	2	1
Aircraft Inspection	45	40	5
Wheel and Tire Supervision	6	4	2
Wheel and Tire	18	16	2
Aerospace Ground Equipment	75	66	9
Munitions	240	192	48
Totals	1,514	1,357	157

[a] Weapons Supervision (and Fabrication Flight Supervision) show a small increase in manpower requirements under two 36 PAA squadrons, because these supervision requirements increase for a single squadron as PAA increase from 24 to 36, unlike most other supervision requirements.

Observe that, out of this 157-position difference, 18 positions are in the MXG or MOS, 48 are in the AMXS (five of which are in supervisory and support functions) and 91 are in the MXS (ten of which are in supervisory and support functions). Thus, of the 157-position reduction, 33 positions (21 percent of the total) are in supervisory and support functions, while 124 are in "touch labor" maintenance. Out of these 124 positions, there are 48 in Munitions, 15 in Aircraft Section (Crew Chiefs), 14 in Aircraft Structural Maintenance, 13 in Weapons Loaders and 12 in Avionics Specialists, while all other "touch labor" maintenance work centers have a reduction of less than ten positions.

Determining Maintenance Manpower Requirements at Active Associate Units

The F-35 manpower requirements identified in the previous section are based on the requirements of combat-coded squadrons that are deployed in support of their wartime tasking. A key distinction between AC and RC squadrons is that RC units divide their total military manpower into full-time and part-time maintenance positions, whereas there are no part-time positions in AC maintenance units. In general, RC total military manpower is sized according to the requirements of the squadron's wartime tasking (as with the AC), but RC full-time maintenance manpower is sized to support the requirements of the unit's home-station training mission.[56]

As discussed earlier, this analysis assumes that all RC units will be structured as Active Associate units, with a number of AC pilots assigned to the unit. The maintenance manpower at Active Associate units could also be composed of a mix of AC and RC maintainers, but it is not apparent how this mix should be determined. As with pilots, the USAF could take advantage of the more-experienced RC maintenance workforce to assist with training of inexperienced AC maintainers. However, this would force RC maintainers to spend less time on direct production tasks and more time supporting maintenance training, eliminating some of the advantage that has historically enabled the RC to support home-station training more efficiently than AC maintenance units (Drew, 2008).

We identified three alternative strategies for determining the mix of AC and RC maintenance positions at an associate unit. We first focus on an alternative that utilizes only RC maintenance manpower at Active Associate units. At the end of this chapter, we will examine two other alternatives that place AC maintenance manpower in the Active Associate unit.

---

[56] Exceptions to this general rule exist, for example, at ANG units that are resourced to support both an Aerospace Control Alert mission and a training mission while at home-station.

Alternative 1: Utilize only RC maintenance manpower. Under this alternative, the AC would not provide any maintenance manpower to the associate units. Instead, all maintenance positions would be filled by AFRC and ANG maintainers.

We worked with the Logistics and Manpower directorates at both Headquarters, AFRC (AFRC/A4 and AFRC/A1) and Headquarters, National Guard Bureau (NGB/A4 and NGB/A1) to determine the split between full-time and part-time manpower requirements for F-35 units.

Because full-time manpower requirements are based upon the unit's home-station training mission, we utilized the findings from our pilot absorption analysis to identify a UTE rate, defined as the number of sorties flown per PAA per month, and an average sortie duration, defined as number of flying hours per sortie, for Active Associate squadrons. The UTE values varied as a function of the squadron size, as discussed in the pilot absorption analysis. We assumed an average sortie duration of 1.4 in all cases.

Based on these training mission requirements, AFRC evaluated three squadron sizes: 24 PAA, 30 PAA and 36 PAA. In each case, AFRC assumed that the total military manpower would be equal to the manpower calculations presented in Table 4.5, with two additional Drill Status positions added to each alternative to account for AFRC-unique requirements (Group Career Advisor and administrative support). The data provided by AFRC detailing the full-time and part-time maintenance manpower requirements are presented in Table 4.8, classified by Resource Identification Code (RIC). Note that these calculations are based on an assumption that all maintenance positions are filled by AFRC personnel.

Table 4.8. Alternative 1: Total Military Manpower and Full-Time Manpower Requirements for Active Associate AFRC F-35 Maintenance Units

		Total Military Manpower			Full Time
Squadron PAA	UTE	RIC 0020: AFRC Officer	RIC 0120: AFRC Drill Enlisted	Total	RIC 0160/163: AFRC Civilian/Technician
24	13.3	9	540	549	241
30	12.4	9	619	628	270
36	11.6	9	706	715	296

NGB evaluated the training mission requirements for two squadron sizes: 18 PAA and 24 PAA. However, NGB determined that they would identify the total military manpower requirements for F-35 squadrons using their own methodology, instead of using the manpower requirements identified previously in Table 4.5.[57] Table 4.9 presents these NGB-determined requirements for total military manpower at each work center.[58] Note that these calculations are based on an assumption that all maintenance positions are filled by ANG personnel.[59]

---

[57] The NGB-derived manpower requirements included positions in the Metals Tech and Propulsion work centers within the MXS. According to F-35 specific data provided in an email from by ACC/A1, these two work centers will not exist for F-35 maintenance. Thus, we deleted the positions associated with these work centers to obtain the values presented in Table 4.9.

[58] The set of work centers presented here is not identical to those presented in Table 4.5 because the data that we received from NGB were in a slightly different format that did not always make it possible to identify separate supervision requirements for some work centers.

[59] These values suggest that the current maintenance manpower at an 18 PAA ANG F-16 unit, which has approximately 550 drill maintenance positions, could support an increased number of F-35 PAA. The NGB estimates that the total maintenance manpower requirement for a 24 PAA F-35 unit would be 566. Thus, with the current total F-16 manpower, an ANG maintenance unit that supports 18 F-16s could nearly support 24 F-35s. Conversely, ANG F-16 squadrons that transition to the F-35 but maintain 18 PAA would need to absorb a reduction of more than 100 maintenance manpower positions, down to a drill maintenance manpower of 440. This is because the F-35 does not have some maintenance requirements that exist for the F-16, e.g., jet engine intermediate maintenance does not exist for the F-35 although it is required for the F-16.

Table 4.9. NGB-Determined Total Maintenance Military Manpower Requirements for Combat-Coded F-35 Squadrons of 18 PAA and 24 PAA

Work Center	18 PAA	24 PAA
Maintenance Group		
Command Section	4	4
Weapons Standardization	5	5
Quality Assurance	13	13
Maintenance Operations Squadron		
Command Section	2	2
Maintenance Operations Flight Supervision	1	1
Maintenance Analysis	4	4
Engine Management	-	-
Maintenance Operations Center	11	11
Plans, Scheduling & Documentation	6	6
Programs and Resources	1	1
Maintenance Training	3	3
Aircraft Maintenance Squadron		
Command Section	4	4
Aircraft Maintenance Unit	6	6
Aircraft Section Supervision	5	6
Aircraft Section (Crew Chiefs)	80	117
Specialist Supervision	2	2
Avionics Specialists	27	43
Weapons Supervision	3	4
Weapons Loading and Weapons Maintenance	63	89
Support Section	-	-
Maintenance Squadron		
Maintenance Supervision	4	4
Command Section	3	3
Accessories Flight Supervision	1	1
Fuel Systems	9	14
Egress	10	12
Fabrication Flight Supervision	1	1
NDI	9	11
Aircraft Structural Maintenance	53	69
Maintenance Flight Supervision	1	1
Aircraft Inspection	9	9
Wheel and Tire	9	9
Aerospace Ground Equipment	19	23
Munitions	72	88
Total	440	566

Due to differences between the format of the data provided by the NGB and the format used in our calculations, it is difficult to perform work center comparisons between these NGB-determined manpower requirements and the requirements identified previously in Table 4.5. However, we can identify the extent of differences in the manpower requirements at the organization level. Table 4.10 presents these differences.

Table 4.10. Differences Between NGB-Determined and RAND-Determined Total F-35 Maintenance Military Manpower Requirements

Organization	NGB-Determined Manpower Minus RAND-Determined Manpower	
	18 PAA	**24 PAA**
Maintenance Group	-5	-5
Maintenance Operations Squadron	-17	-19
Aircraft Maintenance Squadron	-24	15
Maintenance Squadron	7	28
Total	-39	19

For both MXG and MOS, the NGB-determined manpower requirement was smaller for both squadron sizes considered. The NGB-determined manpower did not include any requirement for the Engine Management work center in the MOS. For the AMXS, the NGB-determined manpower requirement was smaller for the 18 PAA case and larger for the 24 PAA case. The largest discrepancies occurred between the requirements for AMU (NGB-determined value was smaller by 13 positions for both squadron sizes), Aircraft Section (Crew Chiefs) (NGB-determined requirement was larger by eight positions for 18 PAA and larger by 26 positions for 24 PAA) and Weapons Loading and Weapons Maintenance (NGB-determined requirement was larger by ten positions for 18 PAA and larger by 23 positions for 24 PAA). Note that the NGB-determined manpower did not include any requirement for the Support Section work center in the AMXS. For the MXS, the NGB-determined manpower requirement was larger for both the 18 PAA case and the 24 PAA case. The largest discrepancies occurred between the requirements for Aircraft Structural Maintenance (including supervision, NGB-determined value was larger by 18 positions for 18 PAA and larger by 28 positions for 24 PAA) and Aircraft Inspection (including supervision, NGB-determined requirement was smaller by five positions for 18 PAA and smaller by seven positions for 24 PAA).

Observe that the maintenance manpower requirements identified by the NGB demonstrate less economy of scale with respect to squadron size than the requirements identified earlier in Table 4.5. The NGB-determined requirement is smaller than the RAND-determined requirement at 18 PAA, and larger at 24 PAA. Thus, the NGB-determined requirement exhibits smaller benefits associated with reorganizing aircraft into a smaller number of larger-PAA squadrons.

Based upon the training mission requirements, NGB identified the full-time and part-time maintenance manpower requirements presented in Table 4.11.

Table 4.11. Alternative 1: Total Military Manpower and Full-Time Manpower Requirements for Active Associate ANG F-35 Maintenance Units

		Total Military Manpower					Full-time			
Squadron PAA	UTE	RIC 0028: ANG Drill Status Guardsman—Officer	RIC 0034: ANG Active Guard Reserve—Officer	RIC 0128: ANG Drill Status Guardsman—Enlisted	RIC 0148: ANG Active Guard Reserve—Enlisted	Total	RIC 0034	RIC 0148	RIC 0160/0170: ANG Civilian /Technician	Total
18	16.8	6	2	409	23	440	2	23	145	170
24	14.5	6	2	531	27	566	2	27	170	199

Determining Maintenance Manpower Requirements at Classic Associate Units

We assumed that all AC units—whether they were Classic Associate units in CONUS, or OCONUS units composed solely of AC pilots—would be composed of AC maintainers only. We made this assumption to allow the entire Classic Associate maintenance unit to be assumed to be available at the AC deploy-to-dwell ratio. One could identify a mix of AC and RC maintenance manpower, but this would generate a reduction in the Classic Associate unit's availability for steady-state deployment, which we did not consider as a reasonable strategy for the F-35.

Determining Annual Manpower Costs for F-35 Maintenance Units

We utilized data provided by the Air Force Cost Analysis Agency and the Directorate of Cost Analysis, Deputy Assistant Secretary of the Air Force for Cost and Economics, to identify the cost associated with each RIC position, these costs appear in Table 4.12 (USAF, 2012).[60] Our analysis also included a cost differential between AC manpower assigned to CONUS-based units and AC manpower assigned to OCONUS-based units. Based on unpublished RAND research, we assumed that AC manpower assigned to OCONUS-based units incurred an additional cost of $25,000 per manpower position per year, beyond the values presented in Table 4.12, to account for differences in pay and permanent change of station costs between OCONUS and CONUS personnel.

Table 4.12. Annual Manpower Cost by RIC

RIC	Annual Cost ($)
RIC 0020: AFRC Officer	42,934
RIC 0120: AFRC Drill Enlisted	18,527
RIC 0160/163: AFRC Civilian/Technician	88,836
RIC 0028: ANG Drill Status Guardsman—Officer	36,476
RIC 0034: ANG Active Guard Reserve—Officer	173,817
RIC 0128: ANG Drill Status Guardsman—Enlisted	13,721
RIC 0148: ANG Active Guard Reserve—Enlisted	95,210
RIC 0160/0170: ANG Civilian /Technician	92,068
RIC 0004: AC Officer	152,209
RIC 0104: AC Enlisted	76,083

[60] For Active AF, the costs presented are the FY 2012 Total Annual Composite Rate, Total Average for Officers and Enlisted. For AFRC and ANG, the costs presented are the FY 2013 President's Budget-Composite Rates for FY 2012.

We estimated AC maintenance units' officer requirements based upon the number of squadrons assigned to a wing. A one-squadron AC wing earned 14 officer positions, a two-squadron AC wing earned 18 officer positions, a three-squadron AC wing earned 22 officer positions and a four-squadron AC wing earned 26 officer positions. These officer positions were assumed to be included in the maintenance manpower totals presented in Table 4.6. Note that, as with RC units, the number of officer positions was assumed to not vary with respect to squadron size. Using this information, we identify the cost of each squadron under consideration; these costs are presented for Active Associate squadrons in Table 4.13 and for AC wings in Table 4.14.

Table 4.13. Total Annual Manpower Cost for Active Associate F-35 Maintenance Units

Squadron Type	Annual Cost ($)
AFRC, 24 PAA	31,800,000
AFRC, 30 PAA	35,840,000
AFRC, 36 PAA	39,762,000
ANG, 18 PAA	21,718,000
ANG, 24 PAA	26,075,000

Note that the annual cost of an AFRC 24 PAA squadron is 20 percent greater than the annual cost of an ANG 24 PAA squadron. The primary reason for this discrepancy is that NGB estimated a smaller full-time manpower requirement for a 24 PAA squadron (199 positions for ANG, versus 241 positions for AFRC), even though the ANG squadron would be supporting a higher UTE rate (14.5 for ANG versus 13.3 for AFRC).

Table 4.14. Total Annual Manpower Cost for AC F-35 Maintenance Units

	PAA per Squadron		
Number of Squadrons per Wing	24	30	36
CONUS			
1	42,683,000	48,694,000	55,313,000
2	79,964,000	91,909,000	104,615,000
3	116,864,000	134,744,000	154,145,000
4	153,613,000	-	-
OCONUS			
1	56,358,000	64,344,000	73,138,000
2	105,789,000	121,659,000	138,540,000
3	154,714,000	178,469,000	204,245,000

Determining Maintenance Manpower Requirements and Costs Across the Alternative Beddowns

Given the annual manpower costs for each squadron and wing under consideration, we can then identify the total annual maintenance manpower cost across each of the 28 alternative F-35 beddowns identified in Chapter Two. The ACC/CC-approved beddown utilizes 16 ANG squadrons of 18 PAA each; four AFRC squadrons of 24 PAA each; six AC OCONUS wings, with two squadrons of 24 PAA at each; and five AC CONUS wings, with one wing having a single squadron of 24 PAA, one wing having two squadrons of 24 PAA, and three wings having three squadrons of 24 PAA. Applying the total annual manpower costs in Tables 4.13 and 4.14 to this beddown, we determine a total annual manpower cost of $1.583 billion. Table 4.15 presents the total annual manpower cost for each of the 28 beddowns under consideration in this analysis.

Table 4.15. Total Annual Manpower Cost by Beddown Alternative

Beddown	Annual Cost ($)
1A	1,479,639,000
1B	1,448,717,000
1C	1,430,671,000
1D	1,399,748,000
1E	1,404,281,000
1F	1,381,559,000
1G	1,361,038,000
1H	1,338,316,000
1I	1,307,887,000
1J	1,281,698,000
2A	1,582,665,000
2B	1,525,233,000
2C	1,526,412,000
2D	1,490,638,000
2E	1,465,884,000
2F	1,429,188,000
2G	1,436,699,000
2H	1,400,003,000
2I	1,386,858,000
2J	1,344,797,000
3A	1,652,221,000
3B	1,590,017,000
3C	1,629,198,000
3D	1,566,994,000
3E	1,541,713,000
3F	1,498,291,000
3I	1,465,136,000
3J	1,422,469,000

Recall that each of these beddowns has a paired beddown that differs only with respect to the percentage of AC squadrons located in CONUS: Beddown 1A is paired with 1B; 1C with 1D; 1E with 1F; etc. For example, beddowns 1A and 1B both utilize 20 ANG squadrons of 18 PAA each; seven AFRC squadrons of 24 PAA each; and a total of 18 AC squadrons of 24 PAA each, but where 1A has one-half the AC squadrons in CONUS, 1B assumes two-thirds of the AC squadrons are based in CONUS. Across all 28 beddowns considered, the maximum difference in cost between any beddown and its mate is less than 4 percent. Thus, we observe that the CONUS/OCONUS mix does not appear to have a significant effect on the total F-35 maintenance manpower cost.

Figure 4.2 presents the total manpower costs associated with each of the 28 beddown alternatives, with the value for each alternative presented as the percentage difference between its cost and the cost of the baseline ACC/CC-approved beddown. Each marker on the figure corresponds to one paired set of alternatives. Because we observed little cost difference between each paired set, we show only the cost differential associated with the member of each paired set that assumes two-thirds of the AC squadrons are based in CONUS, which is the cheaper of the paired set in all instances.[61]

Observe that squadron size has a significant impact on total annual maintenance manpower costs. As the fraction of combat-coded PAA in the AC is held constant, increasing the squadron size (i.e., moving down and to the left on the figure) can significantly reduce the overall maintenance manpower cost. This is consistent with the manpower economies of scale discussed previously. However, as the squadron size is held constant, increasing the fraction of combat-coded PAA in the AC (e.g., comparing the marker farthest to the left for each colored set of markers) increases the overall cost. This occurs because the RC is able to make use of part-time maintainers, who are much less expensive in a nondeployed steady-state role than AC maintainers.

[61] The only exception to this rule is that we show the total cost for beddown alternative 2A, the ACC/CC-approved beddown, which has one-half of the AC PAA in CONUS, rather than showing beddown 2B, which has two-thirds of the ACC PAA in CONUS.

Figure 4.2. Total Annual Maintenance Manpower Costs, by Beddown Alternative

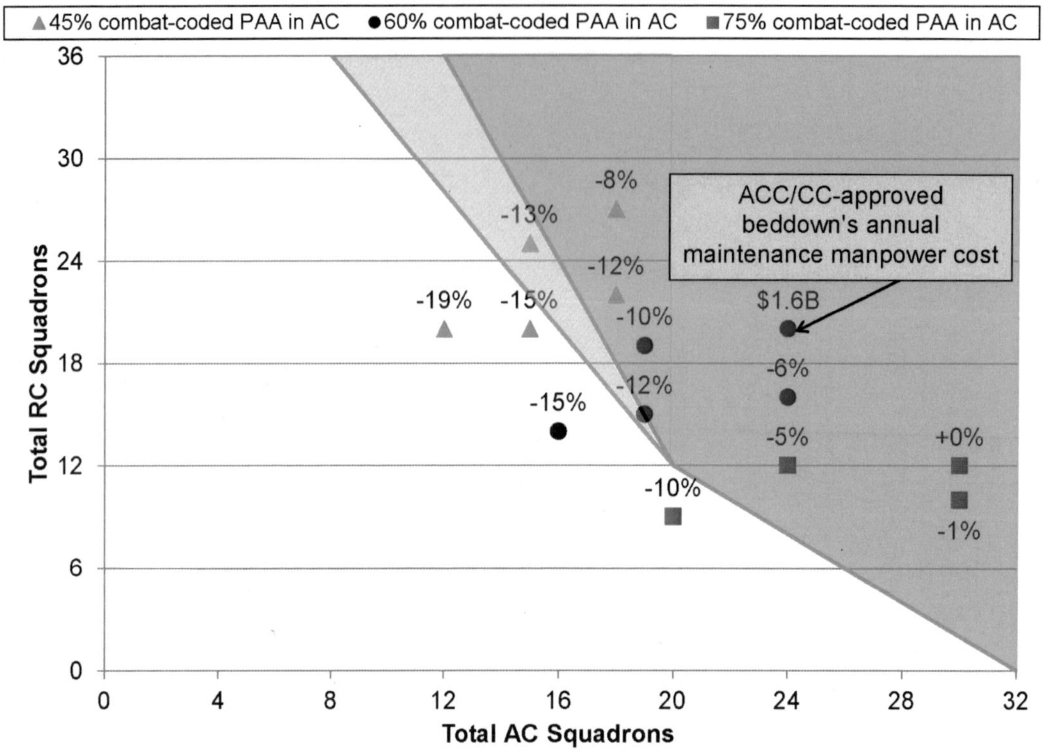

NOTE: The dark green region on this figure corresponds to the range over which all non-surge and post-surge demands can be satisfied within the deploy-to-dwell ratios identified in Table 2.2. The light green region on this figure corresponds to the range over which all rotational requirements can be satisfied within the post-surge deploy-to-dwell ratios.

The minimum cost postures that lie within the dark green region are beddowns 1D, which has 45 percent of the combat-coded PAA in the AC and utilizes squadrons of 24 PAA for all units, and 2H, which has 60 percent of the combat-coded PAA in the AC and utilizes squadrons of 24 PAA for ANG and squadrons of 30 PAA for the AC and AFRC. The annual maintenance manpower cost associated with 1D and 2H is approximately \$183 million (12 percent) less than the cost of ACC/CC-approved beddown. Beddown 2J, which has 60 percent of the combat-coded PAA in the AC and utilizes squadrons of 24 PAA for ANG and squadrons of 36 PAA for the AC and AFRC, can generate an even larger annual cost reduction—approximately \$238 million (15 percent) less than the ACC/CC-approved beddown—but this beddown lies outside of the green regions; it has sufficient squadrons to satisfy surge wartime requirements, but it cannot satisfy steady-state requirements within the desired deploy-to-dwell ratios.

Alternative Strategies for Determining the Composition of AC and RC Maintainers in Active Associate Units

The analysis presented thus far has assumed that Active Associate units utilize only RC maintenance manpower. We next examine two alternative strategies that place AC maintenance manpower in the Active Associate units.

Alternative 2: Increase pilot absorption with AC maintenance manpower. Under this alternative, the AC would provide the maintenance manpower necessary to support increased home-station flying caused by AC pilots.[62] These Active Associate units require a significantly higher UTE rate during home-station training than would a squadron composed entirely of RC pilots, for two primary reasons. First, these AC pilots are less experienced than the RC pilots they are displacing; second, for Active Associations at ANG units, some of the AC pilots are additive to the unit's total pilot population. Thus, it was necessary to determine the full-time maintenance manpower necessary for a unit with only RC pilots, and then identify the additional manpower necessary to support the training requirement of an Active Associate unit with the same PAA.

To identify the full-time maintenance manpower requirements at a unit composed entirely of RC pilots, we identified UTE rates based on squadron size and an assumed average pilot experience level of 80 percent in RC units. As before, we assumed an average sortie duration of 1.4 in all cases.

AFRC and NGB again provided us with estimates of the full-time maintenance manpower requirements for each squadron size under consideration.[63] In every case, the total military manpower values are identical to those presented in Tables 4.8 and 4.11 because the wartime mission requirements are identical across the two alternatives; the only change is to home-station training, which affects full-time manpower requirements only.

Based on the AFRC and NGB data, the full-time manpower requirement at RC units that do not contain AC pilots is only slightly smaller than the full-time manpower requirement at Active Associate units, as presented in Table 4.16.

---

[62] Note that this increased home-station flying was incorporated into the requirements for Alternative 1, but it was not necessary to separate the home-station flying into different segments for Alternative 1, since RC manpower performed all maintenance.

[63] Personal correspondence.

Table 4.16. Full-Time Maintenance Manpower Decrease at RC Units, Were RC Unit Not Supporting AC Pilot Absorption

	UTE Decrease if Unit Does Not Support AC Pilot Absorption		Full-Time Maintenance Manpower Decrease if Unit Does Not Support AC Pilot Absorption	
Squadron Type	Absolute Decrease	Percentage Decrease	Absolute Decrease	Percentage Decrease
AFRC, 24 PAA	1.1	8.3	10	4.1
AFRC, 30 PAA	0.9	7.3	9	3.3
AFRC, 36 PAA	0.7	6.0	10	3.4
ANG, 18 PAA	3.1	18.5	5	2.9
ANG, 24 PAA	2.3	15.9	4	2.0

Both AFRC and NGB evaluated an option with 24 PAA and 12.2 UTE. While AFRC identified a full-time requirement of 231 positions for such a squadron, NGB identified a full-time requirement of only 195 positions. This result is somewhat surprising, since not only are the home-station training requirements equal for each of the AFRC and ANG squadrons, but the ANG squadron has the larger total military manpower (566 positions, versus 549 for the AFRC squadron).

This alternative assumes that the AC will provide the increased maintenance manpower necessary to support the increased home-station flying requirements at Active Associate units. However, because the typical AC maintainer would presumably have less experience than the typical RC maintainer, and because the requirements identified by AFRC and NGB were based on an assumed workforce composed entirely of AFRC and ANG maintainers, AFRC proposed that an equivalency factor of approximately 1.44 AC maintainers per full-time AFRC maintainer be used to identify the AC requirement. Consider, as an example, a 24 PAA AFRC squadron. AFRC identified requirements of 241 full-time AFRC maintenance positions for an Active Associate squadron of 24 PAA, and 231 full-time AFRC maintainers for a 24 PAA squadron composed entirely of AFRC pilots. If we assume that AC maintainers will support the difference in workload between these two squadron alternatives, the ten-position difference for AFRC maintainers would be multiplied by the 1.44 equivalency factor to generate a requirement of 14 AC maintainers at the squadron. We assume that this AFRC-provided equivalency factor also applies to ANG maintainers, and thus apply it to all Active Associate squadrons under consideration, both AFRC and ANG.

As with pilots at associate units, the AC maintainers assigned to Active Associate units could be utilized as either *embedded* or *additive* manpower to the unit. Note that the

AC manpower requirement under Alternative 2 is quite small for each squadron under consideration, never exceeding 14 positions. Were the AC manpower embedded in the unit, it would need to be integrated into the unit's UTCs. AFRC expressed a strong preference for avoiding fragmentation of individual UTCs within any unit due to the difficulties of managing UTCs that are a mix of AFRC and AC personnel. Because these AC maintenance manpower requirements are considerably smaller than any envisioned F-35 maintenance UTC, AFRC determined that it would be preferable to utilize these AC maintainers as additive manpower to the unit. We will thus assume under this alternative that the AC maintenance manpower is additive to all AFRC and ANG Active Associate units, and increases the AFRC or ANG unit's total military manpower.

Tables 4.17 and 4.18 present the maintenance manpower requirements generated by Alternative 2 for AFRC and ANG units, respectively. There are difficulties associated with this alternative: First, although fragmentation of UTCs has been avoided, the unit's total military manpower is now larger than its wartime mission requirement; and second, it might be impractical to assign such a small number of AC maintainers to an Active Associate unit and provide them with the necessary AC supervision and support.

Table 4.17. Alternative 2: Total Military Manpower and Full-Time Manpower Requirements for Active Associate AFRC F-35 Maintenance Units

		Total Military Manpower				Full Time		
Squadron PAA	UTE	RIC 0020: AFRC Officer	RIC 0120: AFRC Drill Enlisted	RIC 0104: AC Enlisted	Total	RIC 0160/163: AFRC Civilian/Technician	RIC 0104	Total
24	13.3	9	540	14	563	231	14	245
30	12.4	9	619	13	641	261	13	274
36	11.6	9	706	14	729	286	14	300

Table 4.18. Alternative 2: Total Military Manpower and Full-Time Manpower Requirements for Active Associate ANG F-35 Maintenance Units

		Total Military Manpower						Full Time				
Squadron PAA	UTE	RIC 0028: ANG Drill Status Guardsman—Officer	RIC 0034: ANG Active Guard Reserve—Officer	RIC 0128: ANG Drill Status Guardsman—Enlisted	RIC 0148: ANG Active Guard Reserve—Enlisted	RIC 0104: AC Enlisted	Total	RIC 0034	RIC 0148	RIC 0160/0170: ANG Civilian/ Technician	RIC 0104	Total
18	16.8	6	2	411	21	7	447	2	21	142	7	172
24	14.5	6	2	531	27	6	572	2	27	166	6	201

NOTE: For an ANG unit with 18 PAA, this five-position reduction included a reduction of two slots in RIC 0148 (ANG Active Guard Reserve—Enlisted) and three in RIC 0160/0170 (ANG Civilian/Technician). This generates a corresponding two-position increase in RIC 0128 (ANG Drill Status Guardsman—Enlisted), because each eliminated RIC 0148 position needs to be offset by an RIC 0128 position to keep the unit's total military manpower at a constant level. For an ANG unit with 24 PAA, this four-position reduction came from four slots in RIC 0160/0170, and thus had no impact on the manpower level in other RICs.

Alternative 3: Increase both pilot absorption and deployment capability with AC maintenance manpower. In addition to increasing pilot absorption capabilities, another potential motivation for the use of Active Associate units is to increase access to RC units for steady-state deployments. In order to generate such an increased deployment capability, one could identify the AC manpower necessary to support an entire set of UTCs, position them within Active Associate units, and make these UTCs available at the deploy-to-dwell ratios assumed for the AC.

As an example, suppose that the AC were to provide the pilots and maintenance manpower equal to the UTC requirement to grow from 12 PAA to 18 PAA at a deployed location. This F-35 UTC requirement equals 70 maintenance positions, as computed by AF/A9 in the Future Forces Database.[64] Under this alternative, the use of *embedded* AC manpower is more practical than it was with Alternative 2, because Alternative 3 avoids fragmentation of UTCs. Thus, we will assume that each AFRC and ANG unit will have its Drill Status maintenance manpower reduced by 70 positions, an amount equal to the Active Associate unit's AC maintenance manpower. As with Alternative 2, AFRC again proposed that equivalency factors be used to adjust the full-time maintenance manpower requirement at Active Associate units. Each Active Associate squadron would have its Drill Status position requirement reduced by 70 positions, but the 70 embedded AC maintainers would be assumed to be capable of generating the full-time maintenance workload of only 48 full-time AFRC maintainers (dividing the total AC maintenance manpower level by the equivalency factor of 1.44), so the AFRC full-time maintenance requirement would be reduced by 48 positions, from 241 to 193 positions. Note that this results in a net gain of 22 full-time positions for each RC squadron, when compared to Alternative 1, but no change to the squadron's total military manpower. We will again apply this same equivalency factor to ANG maintenance requirements.

Tables 4.19 and 4.20 present the maintenance manpower requirements generated by Alternative 3 for AFRC and ANG units, respectively. As was discussed previously, the primary difficulty associated with this alternative is the uncertainty regarding the force presentation concept for the Active Associate units. Note that this analysis assumes that all Active Associate units are structured such that their AC manpower supports a dependent (i.e., "follow-on") UTC capability. This would generate an imbalance between the number of independent (i.e., "lead") UTCs that are available to deploy at AC deploy-to-dwell rates (since there would be one such independent UTC at each AC squadron) and the number of dependent UTCs that are available at AC rates (since there would be one such dependent UTC at each AC squadron and at each Active Associate squadron).

---

[64] Personal correspondence.

Moreover, if the AC portion of an Active Associate squadron deployed but the RC portion remained at home-station, it is not clear if the RC remainder would have sufficient full-time manpower to conduct its home-station mission. This could be overcome by moving some of the Drill Status RC maintainers to a full-time status to fulfill the requirements of the RC remainder's home-station training mission, at some cost.

Table 4.19. Alternative 3: Total Military Manpower and Full-Time Manpower Requirements for Active Associate AFRC F-35 Maintenance Units

		Total Military Manpower				Full Time		
Squadron PAA	UTE	RIC 0020: AFRC Officer	RIC 0120: AFRC Drill Enlisted	RIC 0104: AC Enlisted	Total	RIC 0160/163: AFRC Civilian/Technician	RIC 0104	Total
24	13.3	9	470	70	549	193	70	263
30	12.4	9	549	70	628	222	70	292
36	11.6	9	636	70	715	248	70	318

Table 4.20. Alternative 3: Total Military Manpower and Full-Time Manpower Requirements for Active Associate ANG F-35 Maintenance Units

		Total Military Manpower						Full Time				
Squadron PAA	UTE	RIC 0028: ANG Drill Status Guardsman—Officer	RIC 0034: ANG Active Guard Reserve—Officer	RIC 0128: ANG Drill Status Guardsman—Enlisted	RIC 0148: ANG Active Guard Reserve—Enlisted	RIC 0104: AC Enlisted	Total	RIC 0034	RIC 0148	RIC 0160/0170: ANG Civilian/ Technician	RIC 0104	Total
18	16.8	6	2	339	23	70	440	2	23	97	70	192
24	14.5	6	2	461	27	70	566	2	27	122	70	221

Using this information along with the cost data appearing in Table 4.12, we can identify the cost of each Active Associate squadron under consideration, for each alternative, as presented in Table 4.21.

Observe that, for any squadron type in Table 4.21, the total annual cost differs by no more than 1.3 percent across the three alternatives. Said differently, Alternative 3 provides more deployment capability at essentially the same total cost.[65] However, the number of AC maintainers required at Active Associate units varies significantly across the alternatives, ranging from zero in Alternative 1 to 70 per unit in Alternative 3. Table 4.22 presents the number of AC maintainers required to be assigned at Active Associate units for each of the 28 beddowns under consideration. Note that Table 4.22 presents the AC manpower counts for paired sets of beddowns, since each member of a pair differs only with respect to the percentage of AC squadrons located in CONUS, which has no impact on the AC maintenance manpower requirements at Active Associate units. These AC maintenance manpower positions are in addition to the maintenance manpower positions required at AC units.

Table 4.21. Total Annual Manpower Cost for Active Associate F-35 Maintenance Units, Under Each Manpower Composition Alternative

Squadron Type	Annual Cost ($)
Alternative 1	
AFRC, 24 PAA	31,800,000
AFRC, 30 PAA	35,840,000
AFRC, 36 PAA	39,762,000
ANG, 18 PAA	21,718,000
ANG, 24 PAA	26,075,000
Alternative 2	
AFRC, 24 PAA	31,977,000
AFRC, 30 PAA	36,030,000
AFRC, 36 PAA	39,939,000
ANG, 18 PAA	21,811,000
ANG, 24 PAA	26,163,000
Alternative 3	
AFRC, 24 PAA	31,565,000
AFRC, 30 PAA	35,605,000
AFRC, 36 PAA	39,527,000
ANG, 18 PAA	21,664,000
ANG, 24 PAA	26,021,000

65 Note that the AC manpower was utilized to provide dependent, and not independent, UTCs in this alternative.

Table 4.22. Total AC Maintenance Manpower Required at Active Associate Units, by Beddown Alternative

	Manpower Composition Alternative		
Beddown	**1**	**2**	**3**
1A, 1B	0	238	1,890
1C, 1D	0	180	1,540
1E, 1F	0	205	1,750
1G, 1H	0	155	1,400
1I, 1J	0	152	1,400
2A, 2B	0	168	1,400
2C, 2D	0	128	1,120
2E, 2F	0	157	1,330
2G, 2H	0	125	1,050
2I, 2J	0	116	980
3A, 3B	0	112	840
3C, 3D	0	84	700
3E, 3F	0	96	840
3I, 3J	0	70	630

Observe that, for any beddown, there is significant variation across the three manpower composition alternatives for Active Associate units. For those beddowns that maintain 60 percent of combat-coded PAA in the AC (as in the ACC/CC-approved beddown 2A), Alternatives 1 and 2 can satisfy the AC pilot absorption requirements with between zero and 168 total AC maintenance positions at Active Associate units, while Alternative 3 also provides an increased steady-state deployment capability with between 980 and 1,400 total AC maintenance positions at Active Associate units. Because Table 4.21 demonstrated little difference between the manpower composition alternatives with respect to total annual maintenance manpower costs, the key tradeoff to be considered when evaluating these alternatives is the increase in deployment capability that can be achieved under Alternative 3 versus the increased AC maintenance manpower requirements at Active Associate units.

F-35 Support Equipment Requirements

The second logistics resource that we examined was F-35 SE. The JPO provided us with the SE requirements for AF combat-coded wings and squadrons of varying PAA levels.[66] These requirements do not include Autonomic Logistics Information System equipment. Note that multiple squadrons at one location are assumed to share SE that does not deploy, such as cranes. Table 4.23 presents the costs associated with SE procurement for each of the F-35 wing and squadron sizes under consideration.

Figure 4.3 presents these same SE counts, now normalized on the basis of SE requirements per PAA.[67] As with maintenance manpower, we observe that the SE procurement cost per PAA decreases as the number of PAA per squadron increases. A single squadron of 18 PAA would require $2,389,000 in SE per PAA, whereas a squadron of 36 PAA could be supported by $1,656,000 in SE per PAA—31 percent less. Assigning multiple squadrons to a single wing can generate additional efficiencies beyond those generated by the squadron size effect. While a single squadron of 36 PAA requires $1,656,000 in SE per PAA, a wing of three 36 PAA squadrons requires only $1,467,000 SE per PAA—an 11 percent reduction. Taken together, these effects suggest that increasing the F-35 squadron size and consolidating squadrons into multisquadron wings could generate significant efficiencies in SE procurement requirements. However, unlike the maintenance manpower savings, which are realized annually, these SE procurement reductions are one-time savings.

Table 4.23. Support Equipment Requirements for F-35 Combat-Coded Wings

	SE Procurement (in millions of dollars)			
Number of Squadrons per Wing	18 PAA per Squadron	24 PAA per Squadron	30 PAA per Squadron	36 PAA per Squadron
1	43.0	47.3	55.1	59.6
2	n.a.	84.4	100.0	109.0
3	n.a.	121.5	144.9	158.4
4	n.a.	158.6	n.a.	n.a.

66 Personal correspondence.

67 The single option with four squadrons per wing is not graphed in Figure 2.3.

Figure 4.3. F-35 Support Equipment Requirements per PAA

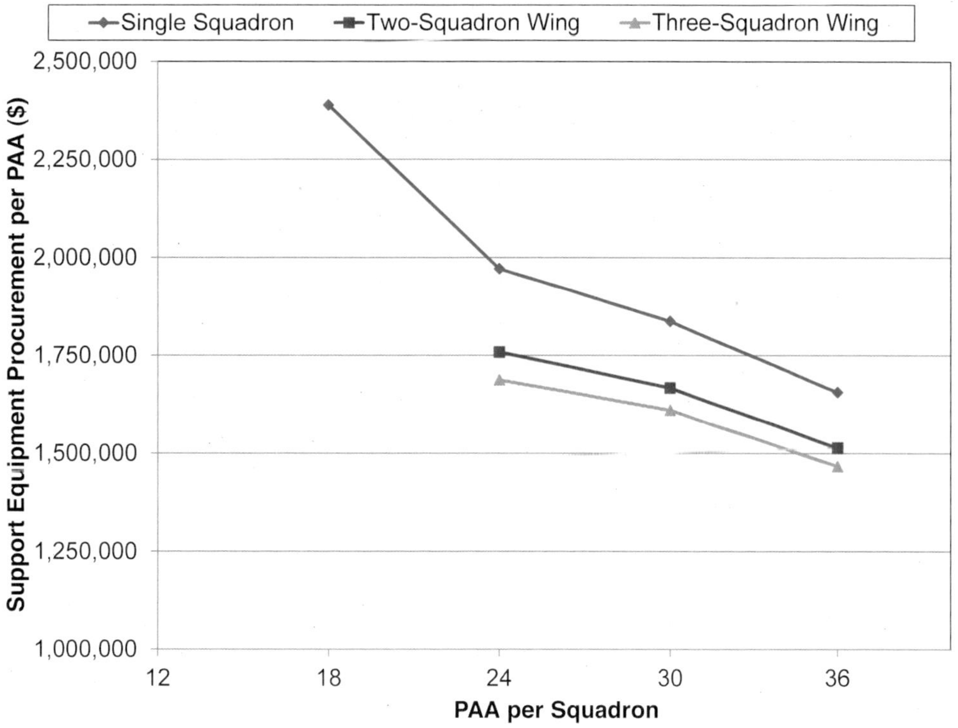

Determining Support Equipment Costs Across the Alternative Beddowns

Given the SE procurement costs for each squadron and wing under consideration, we can then identify the total SE procurement cost across each of the 28 alternative F-35 beddowns identified in Chapter Two. The ACC/CC-approved beddown has a total SE procurement cost of $1.88 billion. Table 4.24 presents the total SE procurement cost for each of the 28 beddowns under consideration in this analysis.

As was discussed in the maintenance manpower analysis, each of these beddowns has a paired beddown that differs only with respect to the percentage of AC squadrons located in CONUS: Beddown 1A is paired with 1B; 1C with 1D; 1E with 1F; etc. Across all 28 beddowns considered, the maximum difference in cost between any beddown and its mate is less than 2 percent. Thus, we observe that the CONUS/OCONUS mix does not appear to have a significant effect on the total F-35 SE procurement cost.

Table 4.24. Total SE Procurement Cost by Beddown Alternative

Beddown	SE Procurement Cost ($)
1A	1,930,300,000
1B	1,940,500,000
1C	1,779,800,000
1D	1,790,000,000
1E	1,880,400,000
1F	1,890,600,000
1G	1,729,900,000
1H	1,740,100,000
1I	1,649,200,000
1J	1,659,400,000
2A	1,879,800,000
2B	1,869,600,000
2C	1,728,800,000
2D	1,749,200,000
2E	1,800,100,000
2F	1,810,300,000
2G	1,683,200,000
2H	1,693,400,000
2I	1,573,200,000
2J	1,583,400,000
3A	1,738,000,000
3B	1,738,000,000
3C	1,677,800,000
3D	1,677,800,000
3E	1,699,400,000
3F	1,719,800,000
3I	1,519,900,000
3J	1,530,100,000

Figure 4.4 presents the total SE procurement costs associated with each of the 28 beddown alternatives, with the value for each alternative presented as the percentage difference between its cost and the cost of the baseline ACC/CC-approved beddown. As with maintenance manpower, each marker on the figure corresponds to one paired set of beddown alternatives. Because we observed that there is little cost difference between each paired set, we show only the cost differential associated with the member of each set that assumes one-half of the AC squadrons are based in CONUS.

Observe that squadron size has a significant impact on total SE procurement costs. As the fraction of combat-coded PAA in the AC is held constant, increasing the squadron size (i.e., moving down and to the left on the figure) can significantly reduce the overall SE procurement cost, demonstrating the economies of scale discussed previously. Furthermore, as the squadron size is held constant, increasing the fraction of combat-coded PAA in the AC (e.g., comparing the marker farthest to the left for each colored set of markers) also decreases the overall cost. This occurs because the ANG is limited to smaller squadron sizes, and when the fraction of total PAA in the AC is increased, fewer PAA are assigned to the smaller-PAA ANG squadrons.

Figure 4.4. Total SE Procurement Costs by Beddown

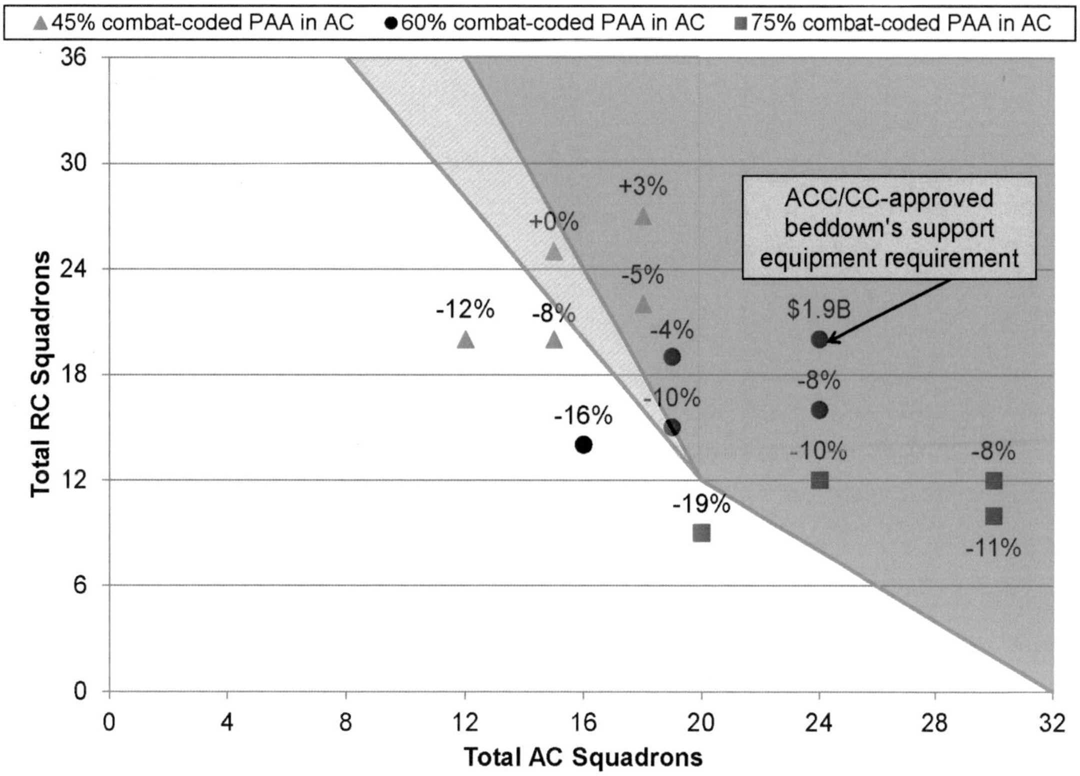

NOTE: The dark green region on this figure corresponds to the range over which all non-surge and post-surge demands can be satisfied within the deploy-to-dwell ratios identified in Table 2.2. The light green region on this figure corresponds to the range over which all rotational requirements can be satisfied within the post-surge deploy-to-dwell ratios.

The minimum cost posture that lies within the dark green region is beddown 3C, which has 75 percent of the combat-coded PAA in the AC and utilizes squadrons of 24 PAA for all ANG, AFRC and AC squadrons. The SE procurement cost associated with beddown 3C is $202 million (11 percent) less than the cost of ACC/CC-approved beddown. Beddown 3I, which has 75 percent of the combat-coded PAA in the AC and utilizes squadrons of 24 PAA for ANG and squadrons of 36 PAA for the AC and AFRC, can generate an even larger cost reduction—$360 million (19 percent) less than the ACC/CC-approved beddown—but this option lies outside of the green regions; it has sufficient squadrons to satisfy surge wartime requirements, but it cannot satisfy steady-state requirements within the desired deploy-to-dwell ratios.

Split Operations Support Equipment Requirements

SE is a resource that often limits a squadron's ability to operate simultaneously out of more than one location. Unlike maintenance manpower, where capacity can be expanded by means such as mandating overtime for maintenance personnel, a squadron has on its table of allowances many pieces of "single-item" SE that constrain operations at a secondary site. Rather than taking the SE reductions identified in Figure 4.4 as cost savings, one could use some of this cost reduction to provide a split-operations SE capability at some combat-coded F-35 squadrons.

The F-35 JPO identified a split-operations SE package that would allow a squadron to deploy six PAA to a secondary site and conduct operations at that site while the remainder of the squadron continued its flying mission at a home-station. To compute this split-operations package, they first identified the SE requirement for an initial squadron of six PAA operating at a location. They next determined the SE requirement for a second squadron of six PAA operating at the same site, and used that as the basis for a split-operations SE package. The rationale for using this value, rather than the requirement for the first squadron of six PAA at a deployed location, is that the requirement for the first squadron at a location includes some SE, such as a floor crane, that is earned on a per-site basis and that would probably not be deployed for a split-operations type of contingency. The JPO computed the procurement cost of such a split-operations SE package for six PAA as $20,900,000.

We then compute the total SE requirement for a split-operations-equipped squadron as follows. As an example, consider a 24 PAA squadron: For each SE item, its requirement is equal to the larger of (a) the 24 PAA requirement, or (b) the 18 PAA requirement plus the 6 PAA split-operations requirement. Similar logic is used for the other squadron sizes under consideration. Table 4.25 presents the SE procurement cost for each of the F-35 wing and squadron sizes under consideration. For each of the wing and squadron sizes examined, the SE procurement cost for a split-operations-equipped squadron is between 24 and 42 percent greater than the SE procurement cost for a squadron without the split-operations SE.

Returning now to the SE procurement cost reductions that were identified in Figure 4.4, we can identify the number of squadrons for which a split-operations SE package could be procured at a total SE procurement cost equal to that of the ACC/CC-approved beddown. Because AC squadrons are assumed to be available more often for steady-state deployments, we will identify the number of split-operations-equipped squadrons utilizing the AC squadron size assumed for each beddown. These values are presented in Table 4.26.[68] This increased capability with split-operations-equipped squadrons, at a

---

[68] Note that for a number of beddowns (1I, 1J, 2I, 2J, 3I, 3J), at a total SE procurement cost equal to that of the ACC/CC-approved beddown's baseline, the number of squadrons that could receive split-operations SE packages is greater than the total number of AC combat-coded squadrons.

constant total cost, is generated by increasing the PAA per squadron and harvesting the resultant economies of scale.

Table 4.25. Split-Operations Support Equipment Requirements for F-35 Combat-Coded Wings

	SE Procurement ($M)			
Number of Squadrons per Wing	18 PAA per Squadron	24 PAA per Squadron	30 PAA per Squadron	36 PAA per Squadron
1	56.0	63.9	68.2	76.0
2	n.a.	117.6	126.2	141.8
3	n.a.	171.3	184.2	207.6
4	n.a.	225.0	n.a.	n.a.

Table 4.26. Number of Squadrons That Could Be Split-Operations-Equipped, at Baseline ACC/CC-Approved Beddown's Cost

Beddown	Number of Split-Operations-Equipped Squadrons
1A	0
1B	0
1C	6
1D	5
1E	0
1F	0
1G	11
1H	10
1I	14
1J	13
2A	0
2B	0
2C	9
2D	7
2E	6
2F	5
2G	15
2H	14
2I	18
2J	18
3A	8
3B	8
3C	12
3D	12
3E	13
3F	12
3I	21
3J	21

Challenges to Realizing Potential Logistics Resource Reductions

We acknowledge that much of the data used in this analysis to estimate F-35 requirements is based on either JPO engineering estimates or the assumption that legacy fighters such as the F-16 or F-22 provide a reasonable proxy for F-35 logistics system performance. While the specific data values used in this analysis are likely to deviate from actual experience once the F-35 is fielded, we expect that the nature of the relationship between squadron size and logistics resource requirements will be maintained. The existence of strong economies of scale between the PAA per squadron and maintenance manpower and SE requirements will continue for the F-35, since the underlying causes of these scale economies (minimum crew sizes in work centers, single-item SE) will not fundamentally change. That is, as actual F-35 logistics parameters and cost data vary from these estimated values, future experience can be expected to deviate from the absolute cost reductions identified in this analysis, but we expect that the relative percentage savings estimated here will be accurate.

5. Infrastructure

Our analysis of infrastructure requirements addressed the capacity of a subset of existing USAF bases to support F-35 squadrons. We examined only the 37 bases that currently support F-16 and A-10 squadrons, either combat- or training-coded, as these are the aircraft the F-35 is designed to replace. This enabled us to compare beddown costs between these current and future fighter aircraft, based upon assumptions about the suitability of certain resources to support tactical fighters. The bases that we considered are listed in Table 5.1.

To assess the capacity of a base to support one or more F-35 squadrons, we sought to compare the resources required to support those squadrons (the demand) to the resources available at that base (the supply). We considered the following six infrastructure resource categories, identified in ACC (2009) as critical: runway, ramp, maintenance hangars, corrosion control hangars, ammunition storage units, and squadron operations and aircraft maintenance unit (AMU) facilities.[69] We did not include simulator facility requirements, which were addressed in the 2009 ACC report, due to uncertainties regarding the simulator requirement at each operational base, in light of ongoing discussions regarding the potential regionalization of F-35 simulators.

Initially, we also investigated base operations support resource categories (e.g., housing, child care, medical facilities, fitness centers) that were addressed in the 2009 ACC report, but we ultimately excluded these from the cost and capacity analysis because, for the AFRC and ANG bases, these requirements apply to the entire community supporting the base rather than to the base itself, and therefore any comparisons we might have made between AC and RC bases would have been on a decidedly unequal footing. A detailed study of each base community would be needed to determine whether any meaningful base operations support shortfalls exist.

---

[69] Note that this is not an exhaustive list of the additional infrastructure that would be required at a current F-16 or A-10 base in order for the base to support F-35 operations. As an example, based on the increased security classification requirements for fifth-generation fighter aircraft, increased costs would be necessary to support a higher level of classification for communications lines, sensitive compartmented information facilities, etc.

Table 5.1. Current F-16 and A-10 Bases

Class	State/Country	Base
AC / CONUS	AZ	Davis-Monthan Air Force Base
AC / CONUS	FL	Eglin Air Force Base
AC / CONUS	UT	Hill Air Force Base
AC / CONUS	AZ	Luke Air Force Base
AC / CONUS	GA	Moody Air Force Base
AC / CONUS	NV	Nellis Air Force Base
AC / CONUS	SC	Shaw Air Force Base
AC / OCONUS	Italy	Aviano Air Base
AC / OCONUS	AK	Eielson Air Force Base
AC / OCONUS	South Korea	Kunsan Air Base
AC / OCONUS	Japan	Misawa Air Bases
AC / OCONUS	South Korea	Osan Air Base
AC / OCONUS	Germany	Spangdahlem Air Base
AFRC	LA	Barksdale Air Force Base
AFRC	FL	Homestead Air Reserve Base
AFRC	TX	NAS JRB Fort Worth (Carswell Field)
AFRC	MO	Whiteman Air Force Base
ANG	NJ	Atlantic City International Airport
ANG	CO	Buckley Air Force Base
ANG	VT	Burlington International Airport
ANG	AL	Dannelly Field (Montgomery)
ANG	IA	Des Moines International Airport
ANG	MN	Duluth International Airport
ANG	AR	Fort Smith Regional Airport
ANG	IN	Fort Wayne International Airport
ANG	CA	Fresno Yosemite Air National Guard Base
ANG	ID	Gowen Field Air National Guard Base (Boise Air Terminal)
ANG	MD	Joint Base Andrews
ANG	TX	Joint Base San Antonio-Lackland
ANG	SD	Joe Foss Field Air National Guard Station
ANG	SC	McEntire Joint National Guard Base
ANG	MI	Selfridge Air National Guard Base
ANG	OH	Toledo Express Airport
ANG	WI	Truax Air National Guard Base (Dane Co-Regional)
ANG	AZ	Tucson Air National Guard Base
ANG	OK	Tulsa International Airport
ANG	MD	Warfield Air National Guard Base (Martin State)

First, we estimated demand requirements for an F-35 squadron (ACC, 2009; Air Force Manual [AFM], 2012; Air Force Reserve Command Handbook [AFRCH], 2012; Air National Guard Handbook [ANGH], 2012; USAF, 2006). To estimate the available infrastructure supply, we relied chiefly on OSD's extensive Facilities Program Requirements Suite database, which lists the size of each individual facility (e.g., each maintenance hangar) at each base, along with the annualized per-unit sustainment, modernization, and operations costs associated with most of these facilities.[70] Where cost data for a particular facility were not available, we used the average cost for that particular type of facility at that particular base. If there were no other facilities of that type at that base, we used the average per-unit cost for that type of facility at that *class* of base (AC/CONUS, AC/OCONUS, AFRC, or ANG). For both supply and demand, we also benefited tremendously from consultation with subject matter experts from all Air Force components, who refined, corrected, and filled in missing information.

Because the specific requirements for an F-35 squadron are crucial to this analysis, we describe them in some detail in the following sections. An important complication here is that most of the requirements were stated in the documents only in terms of a single, standard 24 PAA squadron. In order to examine the set of 28 alternative beddowns considered in this analysis, we considered 11 different base configurations, as shown in Table 5.2.

Table 5.2. Possible Squadron Configurations for a Base

Configuration	Total PAA	Number of Squadrons	PAA per Squadron	Applicability
1	18	1	18	ANG
2	24	1	24	AC, AFRC, ANG
3	30	1	30	AC, AFRC
4	36	1	36	AC, AFRC
5	48	2	24	AC
6	60	2	30	AC
7	72	2	36	AC
8	72	3	24	AC
9	90	3	30	AC
10	96	4	24	AC[a]
11	108	3	36	AC

[a] CONUS only.

As discussed in Chapter One, we assume that each AFRC and ANG squadron is located at a unique base,[71] and that each AC Wing corresponds to one base. In the following sections, we detail the stated requirements and how we extended them. We also

[70] Note that this database does not address the condition or adequacy of current facilities and infrastructure.

[71] As noted in Chapter One, this assumption is consistent with the current beddown of combat-coded AFRC and ANG fighter/attack squadrons. It is possible that multiple RC squadrons could be assigned to a single wing; however this analysis did not consider such alternatives.

note the six-digit FRPS category codes that we used to determine the available supply of these resources at each base.

Runway and Ramp

The requirement for runways is simple:[72] To enable F-35 operations, a base must have (or have access to) at least one runway that is at least 8000 feet long and 150 feet wide (ACC, 2005, Slides 7, 12, 18; AFM, 2012, Table 2.3; AFRCH, 2012, §4.3; ANGH, 2012, Table 2.1). The requirement is the same for all 11 configurations in Table 5.2. For training purposes, the ACC research (2009) indicated that a second runway would be needed if there were at least three squadrons, but for this analysis we applied only the standard operational requirement.[73]

Although most of the ANG bases we considered do not technically own suitable runways, they often share or have access to civilian runways, and ANG provided their dimensions. In the end, we found that all of the bases we considered have access to at least one suitable runway and therefore all 37 met this requirement. Since all these bases currently support F-16s and A-10s, this was not a surprising result.

The requirements for ramp or parking apron capacity are more complex.[74] Fighters are typically parked at an angle, as shown in Figure 5.1. Note that the distance between the rows depends on whether the aircraft are parked nose-to-nose or tail-to-tail, as greater space is needed between the tails for safety due to the hazard of jet engine exhaust.

---

[72] The runway FRPS category code is 111111.

[73] Personal communication from Mr. Thomas Ardern, ACC/A5BA.

[74] The apron capacity FRPS category code is 113321.

Figure 5.1. Angled Parking for Fighter Aircraft

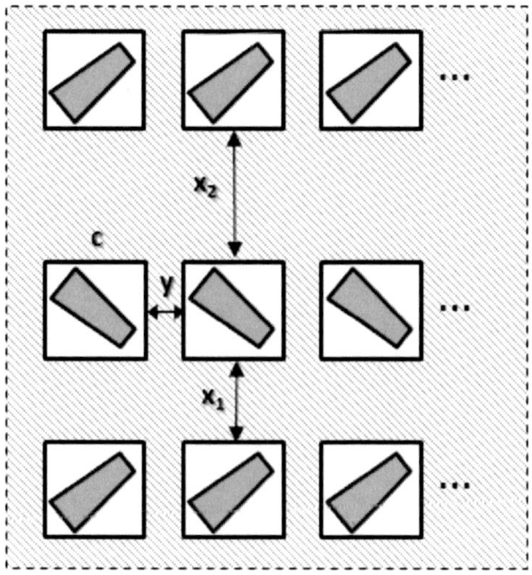

In Table 5.3, we compare the ramp space required for one 24 PAA squadron of F-16, A-10, and F-35 aircraft. Conveniently, the space requirements for the F-35 falls between those of the other two fighters, a fact which indicates that the ramps used for F-16 and A-10 parking are likely to be amenable to F-35 parking as well.[75]

Table 5.3. Angled Parking Comparison for F-35, F-16, and A-10 (One 24 PAA Squadron)

Aircraft	Block (c) (feet)	Gap Between Head-to-Head Rows (x1) (feet)	Gap Between Tail-to-Tail Rows (x2) (feet)	In-Row Spacing (y) (feet)	Total Apron Space Needed (feet × feet)	Area[a] (square yards)
F-35	45	90	135	20	360 × 500	20,000
F-16	40	90	135	21	345 × 467	17,902
A-10	47	97.5	146.25	49	384.75 × 623	26,633

[a] Note the change in units. (Facilities Program Requirements Suite gives ramp area in square yards.)
SOURCE: AFM, 2012, Table 2.11 and Fig. 1; AFRCH, 2012, §4.6; ANGH, 2012,Table 2.5. The ACC research (2009) divides X1 and X2 differently, but shows the same total area.

---

[75] Technically speaking, we enforced only the area requirement, rather than enforcing each dimension separately, because the FRPS database provides only areas. While it is theoretically possible that the parking aprons at some bases may be shaped such that they cannot be used by the F-35A, the fact all of these bases currently support either F-16s or A-10s strongly suggests this should not be an issue.

For configurations involving multiple squadrons, we assumed that each squadron parks separately.[76] For different squadron sizes, we assumed the same basic *3 x n* parking configuration, where *n* could be 6, 8, 10, or 12. If a base had more than one ramp facility listed, we allowed squadrons to be split between ramps if necessary.[77] Here again, we found no shortfalls in terms of ramp space for any configuration at any base.

Maintenance Hangars and Corrosion Control Hangars

The requirements for parking F-35s within a closed maintenance hangar are quite similar to those for parking on an open ramp but are more difficult to compare against existing infrastructure because maintenance hangars are much smaller and an aircraft cannot be split between hangars:[78] A hangar large enough to fit only half an aircraft can, in fact, fit none at all.[79] Moreover, the interior configuration of the hangars can limit how much of the space is actually usable. Fortunately, the maintenance box for the F-35 (the exact footprint within the maintenance hangar that must be reserved for F-35 maintenance) is nearly identical to the maintenance boxes for the F-16 and A-10, which suggests that maintenance hangars that work for one will work for the other.[80]

Figure 5.2 shows that the maintenance box for a single F-35 (ANGH, 2012, Table 7.1) is 5040 square feet. It is worth noting that two aircraft parked next to each other require only ten feet of padding between them (i.e., the shaded areas of the maintenance boxes may overlap), so two aircraft parked together require somewhat less space than if parked separately (about 8 percent less space: 4,640 square feet or 4,725 square feet per

---

[76] The only effect of this more conservative assumption was to require extra "padding" of 20 feet between parked squadrons. In the end, it made no difference to the overall result either way.

[77] We checked that each ramp was large enough to contain the required columns of aircraft. There was no issue in this regard.

[78] The closed maintenance hangar FRPS category code is 211111.

[79] The facilities requirement document for active duty bases (AFM, 2012, §3.1.2.3.1) also permits the maintenance docks 211173, 211175, 211177, and 211179 to be used. However these four docks were excluded from the figures given by ANG subject matter experts, and there is no corresponding indication within ANGH 32-1084 that these codes may be used for ANG bases, so they were not allowed for ANG bases. (Generally speaking, AFRC bases follow active duty requirements unless otherwise specified.) Note that the facilities document does not distinguish low-observable maintenance facility space from other maintenance space for the F-35.

[80] The tightness of the space ordinarily would make the exact configuration more of a concern, but fortunately maintenance hangars are almost always rectangular, and approximate standard spaces are also defined for small, medium, and large aircraft. The F-35A, F-16, and A-10 all use the same type of space.

Figure 5.2. F-35 Maintenance Box

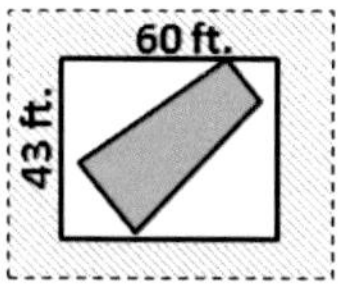

aircraft depending on how they are aligned within the hangar).[81] Therefore, while we require that each hangar must be at least 5,040 square feet to house a single aircraft, we use a slightly smaller figure of 4,700 square feet per aircraft for hangars that could hold more than one. This is the figure used in the ACC report (2009) and is very similar to the average figure of 4,667 square feet required for F-16s and A-10s (ANGH, 2012, Table 7.1).

The maintenance box lays down the requirements for a single aircraft, but it is not necessary for every aircraft in the squadron to have an open berth waiting for it in a maintenance hangar: Only a certain percentage of spaces are needed. This is known as the parking factor (or hangar quantity factor). For an F-35 squadron at an AC base, the parking factor is 27 percent (AFM, 2012, Table 3.1). In the absence of a specific stated requirement for the F-35 for ANG or AFRC bases, we used the AC requirement, in accordance with the general instructions in the corresponding ANG and AFRC facilities documents.[82] Note that the parking factor yields a minimum quantity that must be rounded up to a whole number of spaces—so for example, the 27 percent parking factor for 24 PAA yields 6.48 spaces, which we take to be 7. Note also that we did not apply this parking factor separately to each F-35 squadron, but rather applied it to all of the F-35 aircraft stationed at a base, effectively treating them as part of the same maintenance pool.

---

[81] The size of the modest savings depends on whether the aircraft blocks can be arranged so as to share their 60-foot sides (800 square feet savings), or whether they can only share the 43-foot sides (630 square feet savings). The issue is compounded when three or more aircraft are parked in the same hangar. Of course, these savings assume that the configuration of space inside the hangar is such as to allow side-by-side placement.

[82] "Criteria for items not addressed in this handbook may be found in Air Force Handbook 32-1084, Facility Requirements." (AFRCH, p. 1); "Guidance for criteria not included in this Handbook can be found in Air Force Manual 32-1084, *Facility Requirements*, and must be validated by the Chief, Asset Management Division (NGB/A7A)." (ANGH p. 2). Although the AFRC handbook gives a slightly smaller generic figure of 25 percent for "fighter/helicopter units" (AFRCH, 2012, Table 9.1) the manual mentions the F-35A nowhere by name, so it was not clear to us that this requirement had been considered specifically for the F-35A as well. The sole effect of our more conservative assumption is to require one extra space for the 24, 30, and 36 PAA squadron sizes at these bases.

For the most part, the calculations for space within corrosion control hangars are similar,[83] except that instead of a parking factor, the AC requirement is for one suitable space in a corrosion control hangar per squadron (ACC, 2009, Slide 14). Although the AFRC standard facilities document indicates that corrosion control hangars are currently not authorized for AFRC units as a general rule, we are mindful that the AFRC document never addressed the F-35 directly, and that the corresponding ANG document states that the F-35 has a "unique mission set" in this particular regard (AFRCH, 2012, §9.1; ANGH, 2012, Table 7.10.). Therefore, to be conservative, we chose to apply the requirement to all bases.[84]

Ammunition Storage

A single 24 PAA F-35 squadron requires three above-ground magazine storage facilities, each of 2,500 square feet,[85] and three Hayman igloos, each of 2080 square feet (80 feet x 26 feet) (ACC, 2009, Slide 15).[86] We enforced each requirement separately. For other squadron configurations, we used the same approach as with maintenance hangars, scaling the required storage with the total number of PAA at each base and rounding up to the nearest whole number of appropriate storage units. This is equivalent to assuming that co-located squadrons will share storage space if needed.

Squadron Operations and AMU

The ACC report (2009, Slides 12, 17) provides a single 26,000-square-foot requirement that encompasses both squadron operations facilities and AMU space for one active duty 24 PAA F-35 squadron.[87] This is consistent with the separate figures for these two categories given in the AC facilities requirement document for a 24 PAA squadron (AFM, 2012).[88] AFRC gives a nearly identical breakdown for a 24 PAA fighter squadron,

---

83 The corrosion control hangar FRPS category code is 211159.

84 Note that this requirement is per *squadron*. This means that configuration #7 (3 squadrons of 24 PAA each) will require one space more than configuration #8 (2 squadrons of 36 PAA each) even though both entail the same number of aircraft (72 PAA).

85 The above-ground magazine storage facility FRPS category code is 422258.

86 The Hayman igloo FRPS category code is 422264.

87 The squadron operations facility FRPS category code is 141753. The AMU FRPS category code is 211154.

88 Table 2.19 gives 14,000 square feet for a generic 24 PAA tactical fighter squadron, which it states is based on the previous requirements for 18 PAA F-15 and F-16 squadrons. Although Table 3.5 in this same document gives no explicit figure for F-35A squadrons for AMU space, it gives 11,600 and 9,520 square feet for 18 PAA F-15 and F-16 squadrons respectively for AMU space, which is consistent with the 12,000 square feet that would complete the 26,000 square feet figure from the 2009 ACC report.

and ANG gives smaller, comparable figures for an 18 PAA squadron (AFRCH 2012, Tables 7.4 and 9.5; ANGH, 2012, Tables 5.12 an 7.5.). Table 5.4 presents the requirements that we used in this analysis for an F-35 squadron.[89]

Deciding how to adjust these figures for larger squadron sizes is not obvious, however. Because this resource category includes administrative and recreational spaces (e.g., aircrew lounges, physical fitness rooms) (ANGH, 2012, Table 5.12), we might expect consolidation into larger squadrons to provide some savings: That is, we would not necessarily expect the requirement to scale proportionally with the number of PAA, but we do expect some scaling.

For this resource category, we chose to scale the squadron operations piece and the AMU piece separately. Many of subcategories in the squadron operations portion of the requirement apparently do scale with squadron size (AFRCH, 2012, Table 7.4), and the overall relationship is surprisingly linear but we have no such data for AMU space. Thus we used the relationship in Figure 5.3 to scale the squadron operations piece, but for lack of other data, we retained the AMU space as a fixed, per-squadron quantity.

Table 5.4. Squadron Operations and AMU Space Requirements

Type	Size (PAA)	Squad. Ops. (square feet)	AMU (square feet)	Total (square feet)
AC	24	14000	[12000]	26,000
AFRC	24	19950	6090	[26,040]
ANG	18	13400	8500	[21,900]

NOTE: Derived figures in brackets.

[89] We treat these as per-squadron requirements, in accordance with the AC facilities requirement document (for squadron operations space, see [2] §2.4.14.2; for AMU space see [2] §3.1.6.2). The distinction is irrelevant for ANG and AFRC bases as none of the beddown alternatives contemplate more than one squadron per ANG or AFRC base.

Figure 5.3. Relation of Squadron Operations to PAA

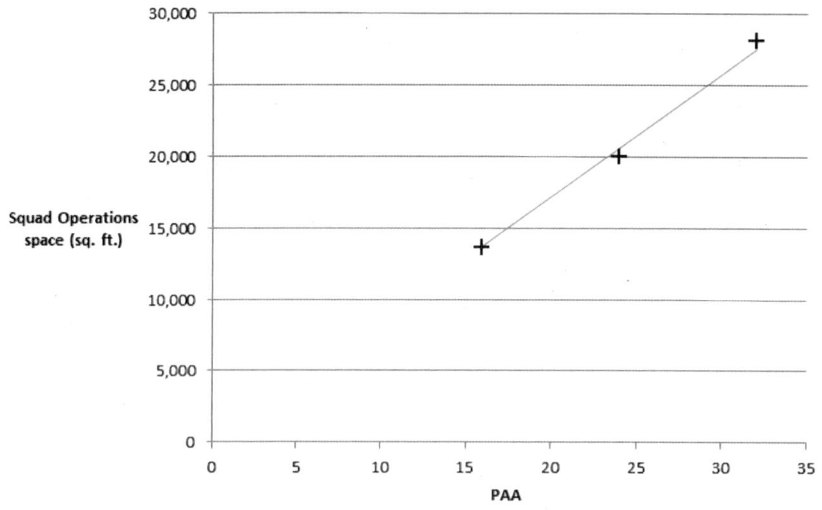

The resulting function for combined squadron operations/AMU space therefore includes some savings for larger squadron sizes. This is a compromise solution that partly scales with PAA and partly does not.[90] Table 5.5 gives the final figures we used for the combined requirement.

[90] Admittedly this is not the most satisfying solution. None of our subject matter experts were able to answer our inquiry for any additional information on this requirement for other squadron sizes; thus this was the best solution we found based on the documents at hand.

Table 5.5. Wing-Level Total Squadron Operations and AMU Space for Each Squadron Configuration

Configuration	Total PAA	Number of Squadrons	PAA per Squadron	Total Square Feet Required		
				AC	AFRC	ANG
1	18	1	18			21,900
2	24	1	24	26,000	26,040	21,900[a]
3	30	1	30	29,500	31,028	
4	36	1	36	33,000	36,015	
5	48	2	24	52,000		
6	60	2	30	59,000		
7	72	2	36	66,000		
8	72	3	24	78,000		
9	90	3	30	88,500		
10	96	4	24	104,000		
11	108	3	36	99,000		

[a] At the request of NGB/A7A, we assumed that an 18 PAA ANG squadron and a 24 PAA ANG squadron would require the same amount of squadron operations and AMU space; namely, the space required for an 18 PAA ANG squadron (ANGH, 2012). We note that this direction is inconsistent with Table 1.5 in the handbook, which states that 18 PAA is the "maximum PAA planning factor used throughout the handbook." When we adjust the ANG requirement based on the process described in this chapter, we estimate a 26,367-square-foot requirement for a 24 PAA ANG squadron. These changes to squadron operations and AMU space of the 24 PAA ANG squadron have a minimal effect on our overall findings (as presented in Figures 5.5, 5.6 and 5.7).

Adjustments for Other Aircraft

To determine the supply of each resource category available at each base, we also need to adjust for the space used to support aircraft other than the F-16 and A-10.[91] Out of the 37 bases listed in Table 5.1, seven bases (Eglin, Moody, Nellis, Barksdale, Andrews, Selfridge and Warfield) have a significant number of fixed-wing, manned aircraft other than the F-16 and A-10. With the exception of Eglin, these bases are assigned only one or two types of other aircraft. For those six bases, we identified the corresponding resource requirements associated with each of these other aircraft, and removed that space from the available supply. In most cases, the adjustments made little or no difference to the final outcome.

We were unable to find sufficient information to make these adjustments for Eglin. Fortunately, it was clear that Eglin has such a large capacity in most of these resource

[91] We assume that all space currently used to support F-16 and A-10 aircraft will be available to support the F-35A. This is consistent with the 2009 ACC report that also explicitly assumed the F-35A as a replacement mission for the F-16 and A-10, for the purposes of computing operating capacity.

categories that adjustments were only relevant for corrosion control hangars and ammunition storage space. For these two categories, we restricted the available space for the F-35 to be only that which is currently used by F-16s and A-10s.

Analysis of Beddown Alternatives

Recall that the beddown alternatives are not site-specific, and do not identify individual bases to be used in any alternative. In order to compare the infrastructure requirements of each beddown alternative with the existing F-16 and A-10 infrastructure, we assigned the configurations to bases so as to make best use of the excess capacity and minimize overall cost. As discussed earlier, some of the resource categories proved to be in sufficient supply for all bases under all beddown alternatives. In particular, for runway and ramp, no new capacity is needed—all F-35 requirements can be satisfied with existing facilities. This is shown in Figure 5.4.

Figure 5.4. F-35 Infrastructure Requirements, Runway and Ramp, by Beddown Alternative

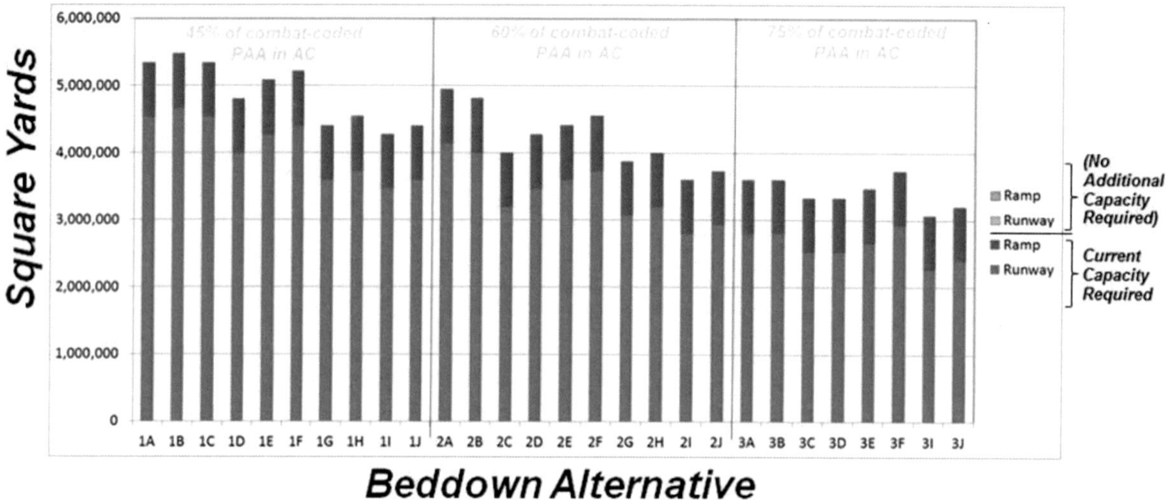

The other resource categories did require some additional capacity in most beddown alternatives. However, as a replacement mission for existing F-16 and A-10 squadrons, the F-35 requires only relatively small amounts of additional capacity. The current and additional capacities required for each alternative are displayed in Figure 5.5. The resource category most commonly in short supply was ammunition storage. Although the two types of ammunition storage are rolled together in Figure 5.5, and there are shortfalls in both areas, the above-ground magazines turn out to be the ones most often lacking.

Figure 5.5. F-35 Infrastructure Requirements, Other Resource Categories, by Beddown Alternative

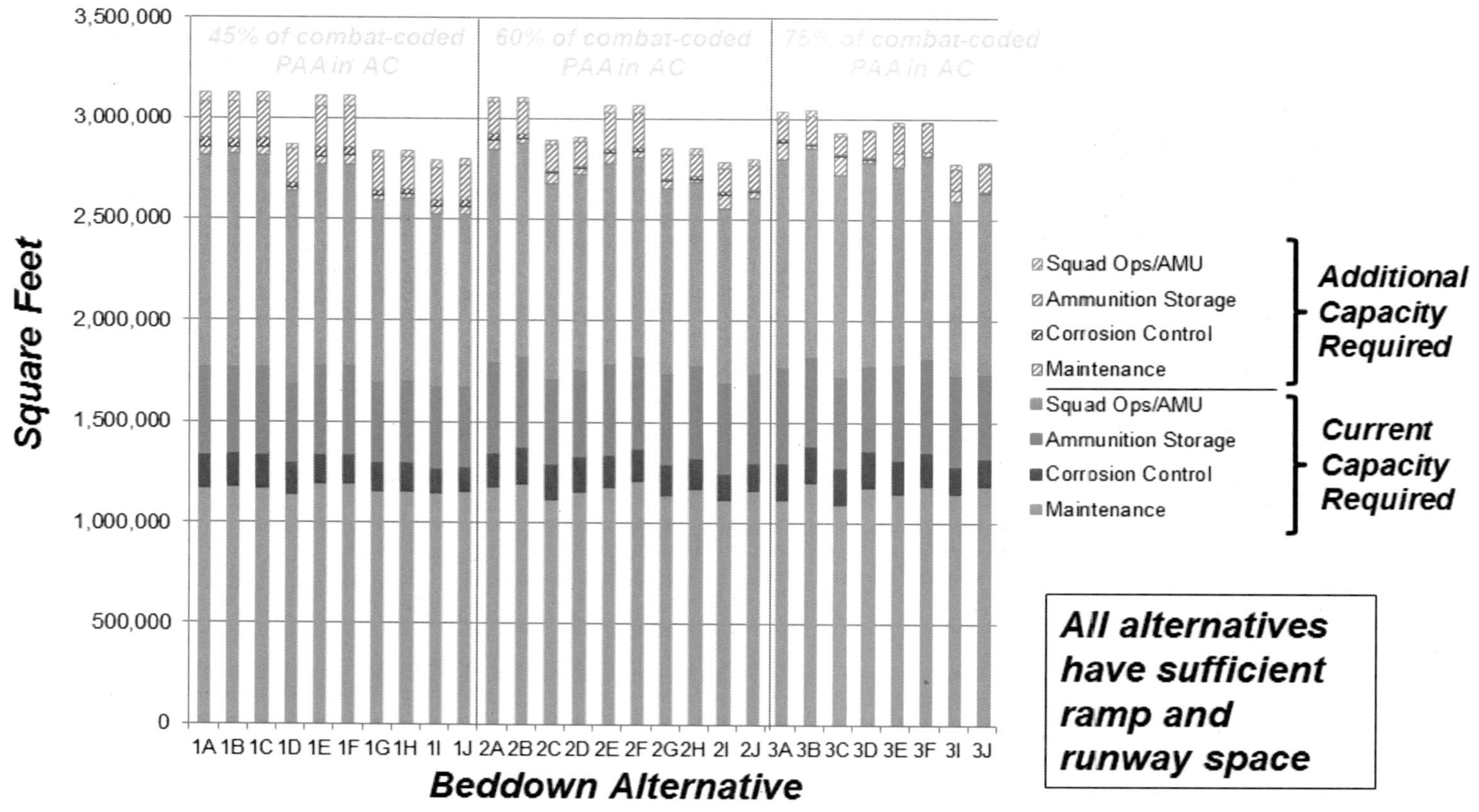

In addition to size and location, the FRPS database also provides per-unit (i.e. per square-foot) cost information for each individual facility. The database gives operations costs, sustainment costs, and modernization costs. We used these figures to estimate the combined current and additional costs that would be required for each beddown alternative. The cost estimates are shown in Figure 5.6. It is worth emphasizing that while there are many infrastructure resource categories, we only considered six basic ones and they should in no way be interpreted as representing total cost. They do, however, reflect relative costs—and given that other resource categories are often tied to the six we chose (for example, maintenance storage facilities are usually linked to the number of maintenance hangars), we expect that these relationships would hold up with additional analysis.

Note that these are all annualized costs. Everything has been amortized across the expected lifetime of the facility, which may be 60 years or more depending on the type of facility. The modernization costs are of particular interest in this regard because they represent the costs required to construct such a facility, when appropriate.[92]

[92] "Modernization may be accomplished through any of the following means: incremental modernization or renewal of a facility over the facility service life; full modernization or renewal of a facility at the end of its service life (such as the Pentagon renovation); and replacement of a facility at the end of its service life by a like facility of equal size." (Office of the Deputy Under Secretary of Defense, Installations an Environment, 2011)

Figure 5.6. Combined Annualized Costs for Each Beddown Alternative

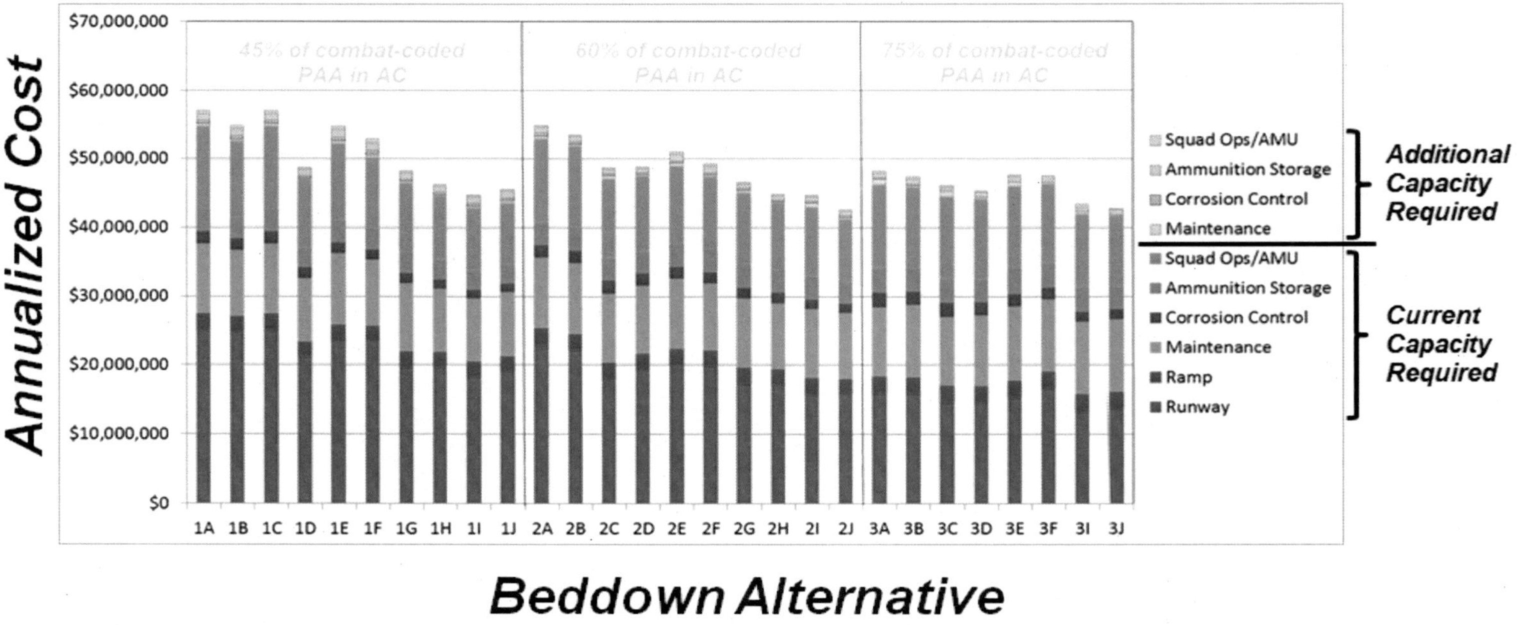

The beddown alternatives exhibit some cost reductions associated with consolidation to fewer bases. Although this is implied in Figure 5.6, these economies of scale are made explicit in Figure 5.7. The figure shows each set of alternatives, distinguished by the percentage of combat-coded PAA in the AC. Beddown alternatives with 45 percent or 60 percent of combat-coded PAA in the AC both exhibit a 21 percent reduction in cost across their range; beddown alternatives with 75 percent of combat-coded PAA in the AC exhibit a 10 percent reduction. As is evident in Figure 5.6, nearly all of the reduction comes from better use of existing capacity.

Figure 5.7. Economies of Scale in Infrastructure Annualized Costs

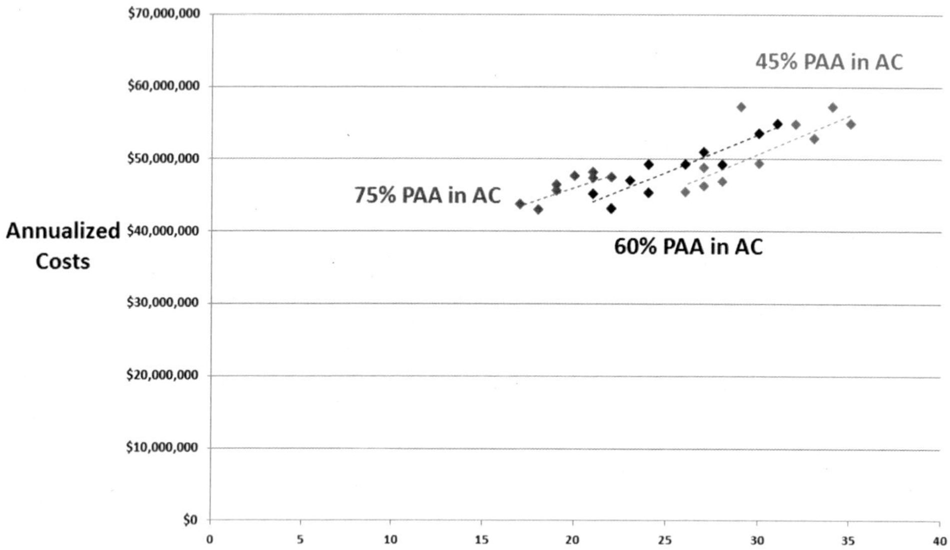

NOTE: The dashed lines are the best-fit least-squares regression estimates for each set of beddown alternatives.

6. Leadership Development

Because the F-35 beddown alternatives would so substantially alter the numbers of F-35s and F-35 units in the active, guard, and reserve components—and, consequently, the numbers of jobs such as squadron commander, group commander, and wing commander that are regarded as key developmental experiences—Air Force decisionmakers wondered whether some alternatives would endanger the development of future senior leaders. To answer the question, we estimated the numbers of positions for fighter pilots under each beddown alternative in each component, identified career paths that develop fighter pilots for different categories of leadership positions, used historical retention behavior and grade distributions to limit possible changes in future fighter pilots' experience and grade profiles, and used two analytic models to assess the Air Force's capacity for developing fighter pilots in the active component under the beddown alternatives.[93]

Numbers of Positions for Fighter Pilots Under the F-35 Beddown Alternatives

About 30 career field managers and corresponding development teams monitor the status of the officer forces in different specialties, assess members' progress and potential, and recommend individuals for in-residence schooling and assignments at higher echelons (e.g., joint, Headquarters Air Force, or major command levels). The rated force has development teams for the Combat Air Forces (CAF), Mobility Air Forces, and special operations forces, rather than for individual weapon systems (e.g., F-35 vs. F-22 vs. C-130). Figure 6.1 displays the conceptual pyramid that guides CAF officer development. The scale on the left lists successively higher grades, from second lieutenant (O-1) through colonel (O-6). At the right are elements of education and training typically pursued at each career stage. The pyramid's face lists major categories of jobs typically held at the various grade levels. This guide does not recommend particular combinations of jobs, differentiate possible career goals, nor distinguish major weapon systems.

[93] Although it was not examined in this analysis, additional work along these lines could illuminate the capacities for developing fighter-pilot leaders in the ANG and AFRC as well.

Figure 6.1. CAF Force Development Pyramid

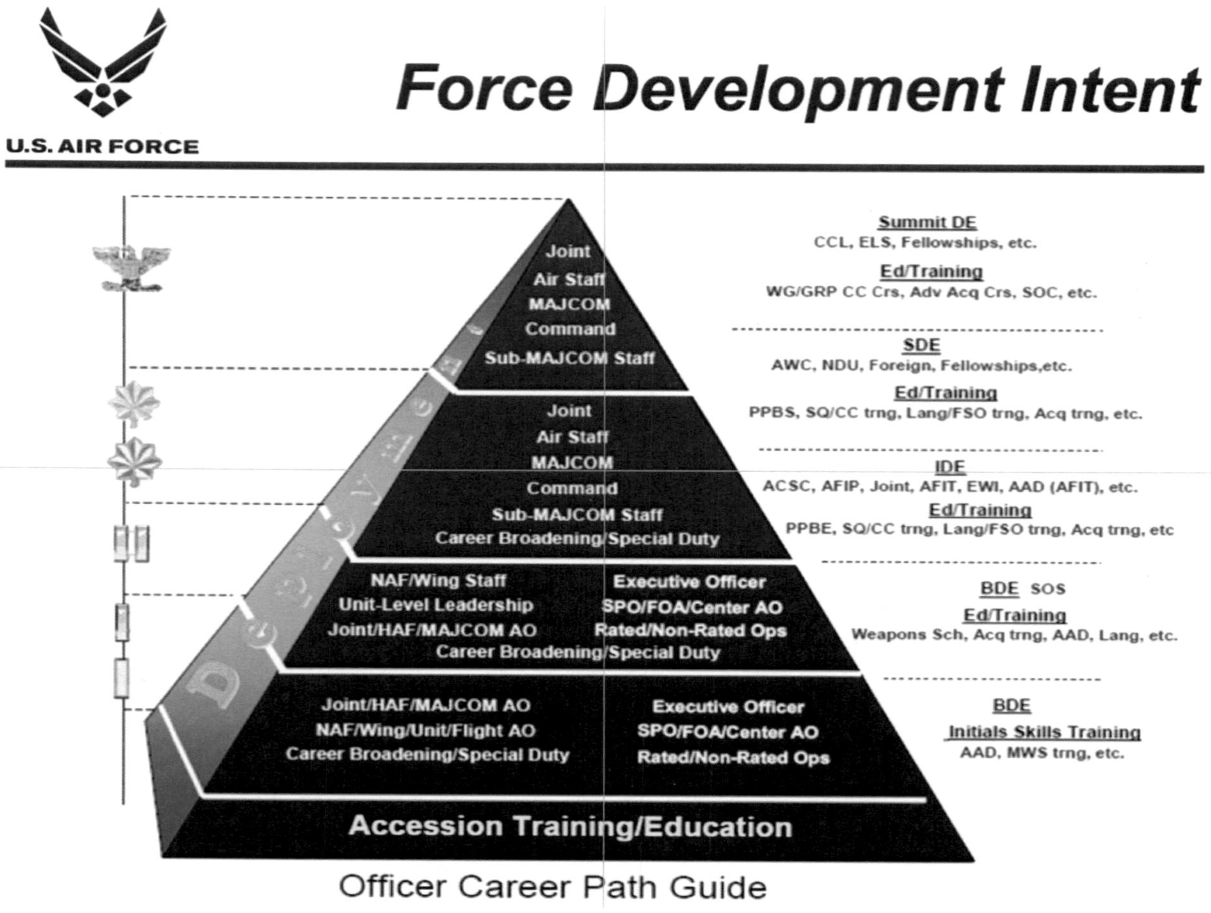

SOURCE: Major David Nuckles, CAF Development Team Advisor, Air Combat Command, June 2012.

Even so, many demands for future senior leaders call for principal experiences in specific major weapon systems—e.g., fighters vs. bombers (Robbert et al., 2005; Moore et al., 2010). Moreover, the CAF development team often divides into fighter, bomber, and Command and Control, Intelligence, Surveillance and Reconnaissance/Electronic Warfare (C2ISR/EW) panels to assess and advise CAF officers. With concurrence from AF/A3O, we elected to assess leader-development capacity at the level of the fighter-pilot force as a whole, not at the coarser CAF level or the finer F-35 level. So we first estimated how the F-35 beddown alternatives would change today's pyramid of active-component jobs for fighter pilots. This involved four broad steps:

1. Determine current authorizations for fighter pilots and other rated officers. We include the latter because multiple specialties share the burden of filling many rated positions that do not necessarily need specific major weapon systems expertise—e.g., instructor (81T) and operations support (16X) billets. Table 6.1 summarizes the rated positions authorized for the active component at the end of FY 2012. Some 2,981 positions called specifically for fighter pilots. Fighter pilots and other rated officers help fill authorizations in the shaded columns left of the "Fighter pilot" column, along with the "Any Combat Support Officer/Air Battle Manager" and "Any Combat Support Officer" columns, accounting for a total of 2,616 additional positions.

2. Estimate the fighter-pilot authorizations at wing level and below for each of the 28 F-35 beddown alternatives. These authorizations depend on the AC/RC mixes, squadron sizes, and unit-association concepts. Our estimates first distributed the API-1 and API-6 strengths for F-35 wings that were calculated for the pilot-absorption analysis in Chapter Three among the relevant job groups and grades using proportions observed in selected operational A-10 and F-16 units, as summarized in Table 6.2.[94] For example, for Alternative 2A's operational F-35 units in the AC, we estimated forcewide totals of:

- 11 O-6 wing commanders
- 22 O-6 serving as wing vice commanders, ops group commanders, or ops group deputy commanders
- 35 O-5 squadron commanders and 29 O-5 ops officers (including both fighter and ops support squadrons)
- 56 O-5 and 55 O-4 authorizations for other wing/Operating Group/Operations Support Squadron jobs

---

[94] For this part of the analysis, we selected only units that were relatively "purely" A-10 and/or F-16 wings, operations groups, operations support squadrons (or flights), and squadrons. For example, although other AC wings in CONUS have operational A-10 or F-16 squadrons, we used only the 20th and 388th Fighter Wings' authorizations to reflect API strength distributions at wing, operations group, and Operations Support Squadron levels.

- 81 O-4 "other" authorizations in fighter squadrons
- 360 O-1/2 and 372 O-3 fighter-pilot authorizations across all organizational levels at wing and below.

This accounts for a total of 1,021 authorizations. Planned unit associations would add 156 O-1/2 authorizations and cut 44 O-3 authorizations at wing and below, and add 20 slots for O-5 squadron commanders (for Active Associate units at ANG and AFRC units), yielding a total estimate of 1,153 authorizations component-wide. Table 6.3 lists the numbers of authorizations estimated across all of the active component's operational F-35 wings under the 28 F-35 beddown alternatives.

Table 6.1. FY2012 Rated Authorizations for the Active Component

Job group		Rated specialty group: Any pilot/navigator/Air Battle Manager	Any pilot/navigator	Fighter pilot/navigator	Other pilot/navigator	Any pilot	Fighter pilot	Other pilot	Any Combat Support Office/Air Battle Manager	Any Combat Support Office	Fighter Combat Support Office	Other Combat Support Office	Air Battle Manager	Grand Total
GO	O-7/O-10		103	3	3		3							112
GO Total			103	3	3		3							112
O-6	O-6 developmental job		3			13								16
	International staff		1			1	8	1						11
	Office of the Joint Chiefs of Staff		1			1	5	3						10
	Combat Command staff		10			6	6	23			2	2	2	51
	Other joint O-6		3			5	5	5		3	2	1	1	25
	Headquarters Air Force staff		7			5	8	5					1	26
	Major command staff		4			17	27	31			1	15	2	97
	Component Numbered Air Force/ Air Operations Center staff O-6		11			13	31	13		1	1	9	1	80
	Wing/Combat Command		25	4	27		7	1						64
	Wing vice commander, Group deputy commander		27	6	39	20	40	37	2			4		175
	Other O-6		9			38	13	7	1	5	1		1	75
O-6 Total			101	10	66	119	150	126	3	9	7	31	8	630
O-4/5	O-4/5 developmental job					20	6	2	11					39
	Joint Staff/combatant command/ international staff					54	90	194	6	29	20	140	73	606
	Other joint O-4/5					51	34	42	5	18	11	34	14	209
	Headquarters Air Force staff					32	61	68	1	11	11	24	15	223
	Major command staff	6				102	87	255	3	22	27	96	30	628
	Component Numbered Air Force/ Air Operations Center staff					63	116	180	1	14	20	97	62	553
	Wing/Group/ Operations Support Squadron staff					82	185	399	3	11	11	108	24	823
	Squadron commander					36	132	193		4	15	46	25	451
	Squadron director of operations					34	109	181		6	9	46	20	405
	Other O-4/5					216	378	1,151	27	54	52	414	174	2,466
O-4/5 Total		6				690	1,198	2,665	57	169	176	1,005	437	6,403
O-1/2	Joint					5	1	24		2		6	56	94
	USAF above wing					96	33	246	9	18	7	111	41	561
	Wing and below					1,105	1,596	5,719	1	110	293	1,609	824	11,257
O-1/2 Total						1,206	1,630	5,989	10	130	300	1,726	921	11,912
Grand Total		6	204	13	69	2,015	2,981	8,780	70	308	483	2,762	1,366	19,057

Fighter pilots and other rated officers help fill authorizations in the shaded columns left of the column "Fighter pilot".

Data source: AF/A3O-AI (Mr. Tom Winslow), March 2012

Table 6.2. Distribution of API Authorizations in Selected A-10/F-16 Units

			AC CONUS				AC OCONUS				ANG				AFRC			
			Wing	Operations group	Operations Support Squadron	Fighter Squadron	Wing	Operations group	Operations Support Squadron	Fighter Squadron	Wing	Operations group	Operations Support Squadron/ Operations Support Flight	Fighter Squadron	Wing	Operations group	Operations Support Squadron/ Operations Support Flight	Fighter Squadron
		Selected no. of A-10/F-16 units	2	2	2	10	4	4	4	10	9	9	9	18	2	2	2	8
		API-6 authorizations	15	9	17	20	29	19	25	20	41	23	49	49	10	8	8	16
		API-3 authorizations													2		1	
		API-1 authorizations		8		283			1	265	7	1	4	399	1		2	170
		Total FY12 authorizations	15	17	17	303	29	19	26	285	48	24	53	448	13	8	11	186
API-6	GO										12%				10%			
	O-6	Wing commander	13%				14%				10%				10%			
		Wing vice commander																
		Group commander/deputy commander	13%	22%			14%	21%			32%	43%			20%	25%		
	O-5	Squadron commander			12%	50%			16%	50%				39%				50%
		Squadron director of operations			6%	50%			4%	50%				37%				44%
		Wing/Group/ Operations Support Squadron staff	27%	44%	12%		24%	21%	12%		32%	43%	55%		60%	63%	50%	
		Other O-4/5												10%				6%
	O-4	Wing/Group/ Operations Support Squadron staff	27%	11%	12%		28%	26%	4%		15%	13%	45%			13%	50%	
		Other O-4/5												14%				
	O-3	Wing and below	20%	22%	59%		21%	32%	64%									
API-3	O-5	Wing/Group/ Operations Support Squadron staff													50%			
	O-4	Wing/Group/ Operations Support Squadron staff													50%		100%	
API-1	O-5	Wing/Group/ Operations Support Squadron staff									29%		25%					
		Other O-4/5												19%				16%
	O-4	Wing/Group/ Operations Support Squadron staff		25%					100%		71%	100%	75%		100%		100%	
		Other O-4/5				12%				9%				81%				84%
	O-3	Wing and below		38%		40%				34%								
	O-1/2	Wing and below		38%		48%				57%								

Source: FY2012 authorizations

NOTE: API-1 positions are line pilots assigned to flying squadrons. API-6 positions are staff or supervisory billets assigned at wing-level or below that require incumbents to fly regularly. API-3 positions are staff or supervisory billets assigned at wing-level or below that require rated expertise but do not require (or in practice, allow) incumbents to fly. See AFI 11-401.

3. Delete the authorizations associated with F-16 and A-10 wings. The F-35 will replace F-16s and A-10s, but the replacement will not be one for one—the total aircraft inventory will decrease. To determine the net change in resulting pilot authorizations, we need to estimate the total number of F-16 and A-10 fighter authorizations at the wing level and below that will be eventually be eliminated. (We did not change fighter positions above the wing level or in units such as A-10 and F-16 training and test units, assuming they would persist as fighter pilot billets even under the new F-35 beddown alternatives.)

We eliminated 20 fighter squadrons from the AC, 18 from ANG, and 8 from AFRC. We also eliminated parts or all of their affiliated wings, operations groups, fighter groups, and operations support squadrons or flights. These amounted to nine AC wings, 18 ANG wings, and three AFRC wings, the latter supporting the eight deleted AFRC fighter squadrons through eight operations or fighter groups and their corresponding operations support squadrons or flights.

The result was that a total of 737 positions were eliminated from the AC:

- 6 O-6 wing commander
- 15 O-6 wing vice or deputy commander positions
- 31 O-5 and 34 O-4 wing/group/ Operations Support Squadron staff positions
- 28 O-5 squadron commanders
- 22 O-5 squadron directors of operations
- 55 O-4 positions
- 256 O-3 positions
- 290 O-1 and O-2 positions.

Subtracting these from the 2,981 fighter pilot positions in Table 6.1 leaves 2,244 authorizations specifically for fighter pilots, plus the 2,616 flexible slots, most of which fighter pilots could/should help fill.

Table 6.3. Estimated Authorizations for AC Fighter Pilots in Operational F-35 Wings

	Wing and below		Other O-4/5	Squadron Director of Operations	Squadron Commander	Wing/group Operations Support Squadron staff		Wing Vice Commander, Group Commander, Group Deputy Commander	Wing Commander	Total
Alternative	O-1/2	O-3	O-4	O-5	O-5	O-4	O-5	O-6	O-6	
1A	470	246	63	21	52	35	36	14	7	944
1B	466	245	65	21	53	40	41	16	8	955
1C	440	238	61	21	47	35	36	14	7	899
1D	427	244	65	21	48	40	41	16	8	910
1E	483	250	64	18	47	35	36	14	7	954
1F	474	257	67	18	48	40	41	16	8	969
1G	443	250	64	18	42	35	36	14	7	909
1H	434	257	67	18	43	40	41	16	8	924
1I	429	238	61	14	38	30	30	12	6	858
1J	417	245	65	15	39	35	36	14	7	873
2A	516	328	81	29	55	55	56	22	11	1,153
2B	500	314	86	28	54	50	52	20	10	1,114
2C	484	303	81	27	48	40	41	16	8	1,048
2D	468	314	86	28	50	50	52	20	10	1,078
2E	514	307	81	22	46	40	41	16	8	1,075
2F	499	312	86	23	47	45	47	18	9	1,086
2G	481	308	81	22	42	40	41	16	8	1,039
2H	466	313	86	23	43	45	47	18	9	1,050
2I	472	306	81	19	37	35	36	14	7	1,007

Alternative	Wing and below		Other O-4/5	Squadron Director of Operations	Squadron Commander	Wing/group Operations Support Squadron staff		Wing Vice Commander, Group Commander, Group Deputy Commander	Wing Commander	Total
	O-1/2	O-3	O-4	O-5	O-5	O-4	O-5	O-6	O-6	
2J	455	314	87	19	38	40	41	16	8	1,018
3A	543	369	101	34	51	45	45	18	9	1,215
3B	522	361	108	34	51	45	47	18	9	1,195
3C	528	368	101	34	49	45	45	18	9	1,197
3D	507	360	108	34	49	45	47	18	9	1,177
3E	560	367	101	27	44	40	41	16	8	1,204
3F	540	380	108	28	46	50	52	20	10	1,234
3I	526	373	101	23	37	40	41	16	8	1,165
3J	509	379	107	24	38	45	47	18	9	1,176

NOTE: Includes AC pilots in Active Associate and Classic Associate units.

4. Add the authorizations associated with F-35 wings. For each beddown alternative, we add the F-35 wing-level and below requirements to the allocation of 2,616 flexible slots to fighter pilots and others based on current authorizations for others and total estimated authorizations for fighter pilots.[95] Table 6.4 summarizes the final results, estimates ranging from 3,764 (Alternative 1I) to 4,189 (Alternative 3F) total AC authorizations for fighter pilots, the latter 11 percent more than the former. Relative differences were notably greater for some key categories of jobs; e.g., wing commander positions for fighter pilots varied by 25 percent, wing vice commander and group commander/deputy commander positions by 17 percent, O-4/5 wing/group/ Operations Support Squadron staff by 26 percent, O-5 squadron/CC by 19 percent, and O-4/5 squadron director of operations by 19 percent. Such large differences should affect the Air Force's capacity for developing fighter pilots into future senior leaders.

---

[95] This process approximated AF/A30-AI's methods: (1) By grade and job group, we allocated 1,290 authorizations for 11K trainer pilots, to different major weapon systems' families of pilots based on 11K suffixes and fighters' share of the active component's total API-1 authorizations. Fighter pilots received shares of these four suffix groups: about 83–86 percent of 276 suffix-D = T-38 jobs that they share with bomber pilots, and about 26–29 percent of 638 suffix-F = T-6 jobs, 36 suffix-Y = general jobs, and 71 suffix-Z = other jobs that they share with bomber, mobility, C2ISR/EW, and special operations pilots. (2) Also by grade and job group, we allocated 951 other flexible authorizations to different major weapon systems' families of pilots based on fighters' share of various groups of rated authorizations. Pilots received about 29–31 percent of 731 positions for any pilot, 87–89 percent of 12 for any fighter pilot or fighter navigator, 22–24 percent of 202 for any pilot or navigator, and 21–23 percent of six for any pilot, navigator, or Air Battle Manager. We left the allocations in fractions (rather than rounding them to whole numbers) to better reflect averages.

Table 6.4. Estimated Total AC Authorizations for Fighter Pilots, Under 28 F-35 Beddown Alternatives

F-35 basing alternative	O-2/3: Wing and below	O-2/3: USAF above wing	O-2/3: Joint	O-2/3 Total	O-4/5: Other O-4/5	O-4/5: Squadron director of operations	O-4/5: Squadron commander	O-4/5: Wing/Group/Operations Support Squadron staff	O-4/5: Component Numbered Air Force/Air Operations Center staff	O-4/5: Major command staff	O-4/5: Headquarters Air Force staff	O-4/5: Other joint O-4/5	O-4/5: Joint Staff/Combatant Command/international staff	O-4/5: O-4/5 developmental job	O-4/5 Total	O-6: Other O-6	O-6: Wing vice commander, Group commander and deputy commander	O-6: Wing commander	O-6: Component Numbered Air Force/Air Operations Centerstaff O-6	O-6: Major command staff	O-6: Headquarters Air Force staff	O-6: Other joint O-6	O-6: Combatant command staff	O-6: Office of the Joint Chiefs of Staff	O-6: International staff	O-6: O-6 developmental job	O-6 Total	GO	Grand Total
1A	2,021	55	2	2,079	454	119	167	217	136	120	71	50	107	12	1,455	38	68	23	43	38	14	9	14	6	9	9	272	55	3,861
1B	2,016	55	2	2,073	457	119	168	227	136	121	71	51	107	12	1,470	38	70	24	43	38	14	9	14	6	9	9	276	56	3,875
1C	1,979	55	2	2,036	452	119	162	217	136	120	71	50	107	12	1,448	38	68	23	43	38	14	9	14	6	9	9	272	55	3,811
1D	1,971	55	2	2,028	457	119	163	227	136	121	71	51	107	12	1,464	38	70	24	43	38	14	9	14	6	9	9	276	56	3,824
1E	2,040	55	2	2,098	455	116	162	217	136	120	71	50	107	12	1,448	38	68	23	43	38	14	9	14	6	9	9	272	55	3,872
1F	2,038	55	2	2,095	459	116	163	227	136	121	71	51	107	12	1,463	38	70	24	43	38	14	9	14	6	9	9	276	56	3,890
1G	1,996	55	2	2,053	455	116	157	217	136	120	71	50	107	12	1,442	38	68	23	43	38	14	9	14	6	9	9	272	55	3,822
1H	1,994	55	2	2,051	459	116	158	227	136	121	71	50	107	12	1,458	38	70	24	43	38	14	9	14	6	9	9	276	56	3,840
1I	1,967	55	2	2,024	451	112	153	206	136	120	71	50	107	12	1,417	38	66	22	43	38	14	9	14	6	9	8	268	55	3,764
1J	1,961	55	2	2,018	456	113	154	217	136	120	71	50	107	12	1,436	38	68	23	43	38	14	9	14	6	9	9	272	55	3,782
2A	2,163	56	2	2,222	475	127	171	258	137	122	72	51	108	13	1,533	39	77	28	44	39	14	9	14	6	9	9	288	57	4,100
2B	2,130	56	2	2,188	480	126	170	249	137	122	71	51	108	13	1,526	39	75	27	44	38	14	9	14	6	9	9	284	57	4,055
2C	2,100	56	2	2,158	474	125	164	228	136	121	71	51	107	12	1,489	38	70	24	43	38	14	9	14	6	9	9	276	56	3,979
2D	2,094	56	2	2,152	480	126	166	249	137	122	71	51	108	13	1,521	39	75	27	44	38	14	9	14	6	9	9	284	57	4,015
2E	2,138	56	2	2,196	473	120	162	228	136	121	71	51	107	12	1,481	38	70	24	43	38	14	9	14	6	9	9	276	56	4,009
2F	2,127	56	2	2,185	479	121	163	239	136	121	71	51	107	12	1,501	38	73	26	44	38	14	9	14	6	9	9	280	56	4,023
2G	2,102	56	2	2,160	473	120	158	228	136	121	71	51	107	12	1,477	38	70	24	43	38	14	9	14	6	9	9	276	56	3,969
2H	2,091	56	2	2,149	479	121	159	239	136	121	71	51	107	12	1,497	38	73	26	44	38	14	9	14	6	9	9	280	56	3,982
2I	2,090	56	2	2,148	472	117	152	217	136	120	71	50	107	12	1,457	38	68	23	43	38	14	9	14	6	9	9	272	55	3,932
2J	2,080	56	2	2,138	479	117	154	228	136	121	71	51	107	12	1,476	38	70	24	43	38	14	9	14	6	9	9	276	56	3,945
3A	2,238	57	2	2,297	495	132	167	237	137	122	72	51	108	13	1,533	38	73	26	44	38	14	9	14	6	9	9	280	56	4,166
3B	2,206	57	2	2,265	502	132	167	239	137	122	72	51	108	13	1,543	38	73	26	44	38	14	9	14	6	9	9	280	56	4,144
3C	2,221	57	2	2,280	495	132	165	237	137	122	72	51	108	13	1,530	38	73	26	44	38	14	9	14	6	9	9	280	56	4,146
3D	2,189	57	2	2,247	502	132	165	239	137	122	72	51	108	13	1,540	38	73	26	44	38	14	9	14	6	9	9	280	56	4,124
3E	2,255	57	2	2,314	494	125	160	228	136	121	71	51	108	12	1,507	38	70	24	43	38	14	9	14	6	9	9	276	56	4,153
3F	2,247	57	2	2,306	502	126	162	249	137	122	72	51	108	13	1,542	39	75	27	44	38	14	9	14	6	9	9	284	57	4,189
3I	2,224	57	2	2,283	493	121	144	228	136	121	71	51	107	12	1,485	38	70	24	43	38	14	9	14	6	9	9	276	56	4,099
3J	2,212	57	2	2,271	500	122	145	239	136	121	71	51	108	12	1,506	38	73	26	44	38	14	9	14	6	9	9	280	56	4,113
Maximum	2,255	57	2	2,314	502	132	171	258	137	122	72	51	108	13	1,543	39	77	28	44	39	14	9	14	6	9	9	288	57	4,189
Minimum	1,961	55	2	2,018	451	112	144	206	136	120	71	50	107	12	1,417	38	66	22	43	38	14	9	14	6	9	8	268	55	3,764
Relative range	15%	5%	6%	15%	11%	19%	19%	26%	1%	2%	1%	2%	1%	3%	9%	3%	17%	25%	1%	1%	2%	2%	3%	1%	1%	5%	8%	5%	11%

Note: Most of the numbers in this table have non-integer values because they represent calculated averages (or expectations), including fighter pilots' proportional shares of flexible authorizations that can be filled by one or more other rated specialties. We display them here as integers, for conciseness.

Developmental Career Paths for Fighter Pilots

Different developmental goals call for somewhat different career paths. For this analysis we focused on well-prepared candidates for selection as general officers along one of two broad developmental paths:

- Joint warfare roles as, e.g., combatant command commanders, vice commanders, operators, and planners
- Organize/train/equip roles as, e.g., chief, vice chief, or assistant chiefs of staff at Air Force headquarters or headquarters in the major commands.[96]

We postulated two other career paths for fighter pilots who could rise to colonel, one oriented toward organize/train/equip roles, and one for general purposes, essentially filling in O-6 jobs as necessary. In addition, we considered two other career paths: one representing officers who advance into the field grades but not to colonel, the other representing officers who leave before promotion to major. Both of these paths fill jobs in their grades that the other four paths do not fill completely.

Figure 6.2 depicts the six career paths and the categories of jobs regarded as mandatory, preferred, and acceptable for each.[97] For example, fighter pilots on the path toward competing for organize/train/equip responsibilities as general officers ideally should have assignments at both major command and Headquarters Air Force levels while they are majors or lieutenant colonels and again while they are colonels.[98] While AC leaders consulted during our development of these paths did not substantially alter this depiction of developmental objectives and career paths, the Air Force's fighter community itself should delineate the appropriate paths, probably working with the CAF development team, AF/General Officer Management Office, and AF/Colonel Management Office (or these offices' successors). During this research the ANG and

---

[96] These developmental approaches had been employed in earlier, unpublished analysis for the Air Force Chief of Staff and the Secretary of the Air Force. If desired, additional research could allow distinct paths to be tailored to grow fighter pilots with more specific areas of expertise—e.g., in acquisition, logistics, human resources, or political-military affairs.

[97] Each path depicted here actually represents a family of paths. Every officer following a given path should hold one or more jobs in each of its mandatory categories, but not necessarily in the preferred or acceptable categories. Different numbers may hold jobs in those categories, and individuals' combinations of experience will differ correspondingly. For example, while they are majors or lieutenant colonels, 50 percent of fighter pilots on the path toward joint warfare general officer may gain experience on a major command staff, while only 10 percent get so-called developmental jobs. (Jobs as 63A acquisition managers or 16P political-military strategists are examples of developmental assignments for fighter pilots.)

[98] It is worth emphasizing that this analysis aims to assess the Air Force's capacity for developing officers with targeted combinations of experience, not to establish check-box lists of jobs that officers might seek with the expectation of reaching the associated career pinnacles.

AFRC began revising the job categories, career paths, and experience priorities in Figure 6.2 to better reflect their organizational structures, job mixes, and developmental objectives—e.g., the ANG eliminated the joint categories of jobs, added ANG state headquarters, and eliminated the path leading toward joint-warfare general officers—so that their capacities for developing senior leaders also could be assessed, but it would take more time and effort bring their assessments to the same level as that produced in this analysis for AC fighter pilots.[99]

---

[99] RAND carried the analysis through for ANG and USAFR fighter pilots using preliminary assumptions about employment categories, job categories, paths, and priorities, etc., for illustrative purposes, but considerable review and revision of assumptions and data would be needed before credible assessments could be drawn.

Figure 6.2. Some Candidate Career Paths for AC Fighter Pilots

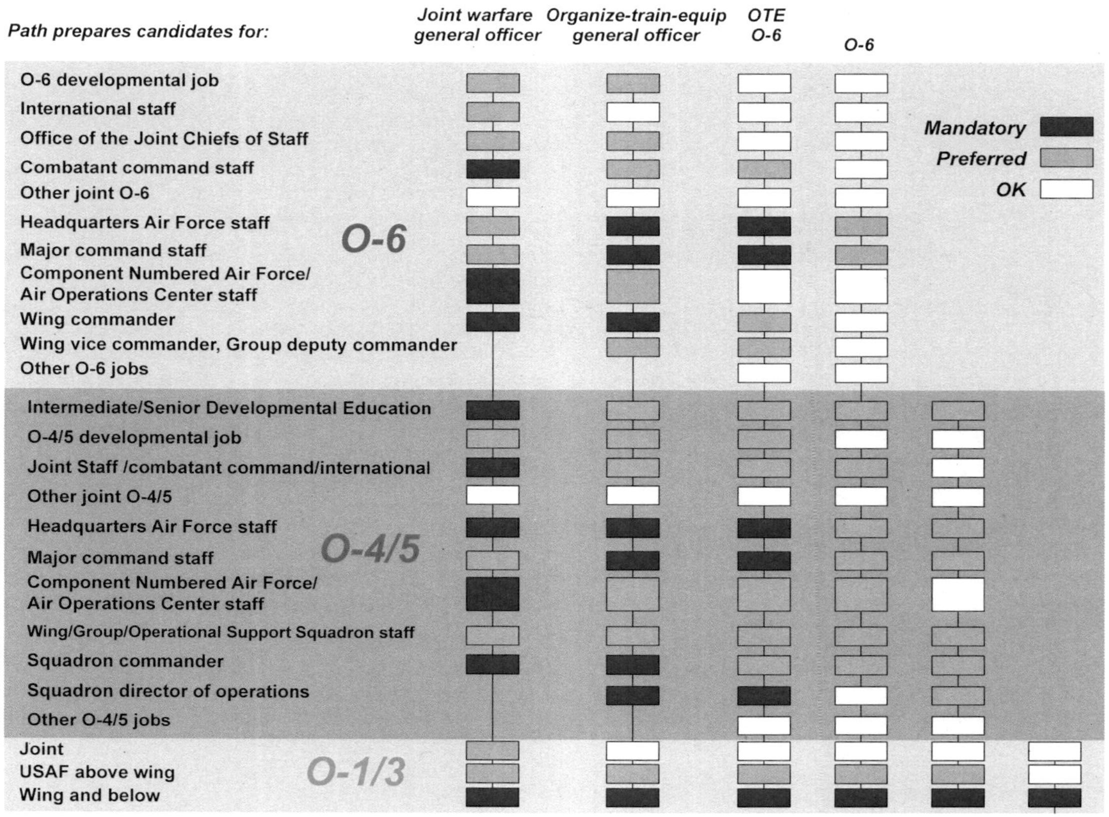

Retention Behavior and Grade Distributions

To help guide and control the analytic flows of fighter pilots through the jobs and career paths now identified, we used historical data to derive the steady-state experience/grade profile for fighter pilots displayed in Figure 6.3. The heights of the graph's bars represent the declining fractions of an entering cohort of fighter pilots expected to remain in the Air Force through increasing numbers of years of commissioned service, calculated as the product of the average continuation rates (during 2000-2011) for that and previous years of commissioned service.[100] The grade breakdown within each bar reflects the average fraction of the cohorts in each grade during that year of service during 2000-2011. For example, about half of an entering cohort of fighter pilots would be expected to remain through their 15th year of commissioned service; most would be majors then, but two years later most would have been promoted to lieutenant colonel.

Figure 6.3. Steady-State Experience/Grade Profile for AC Fighter Pilots

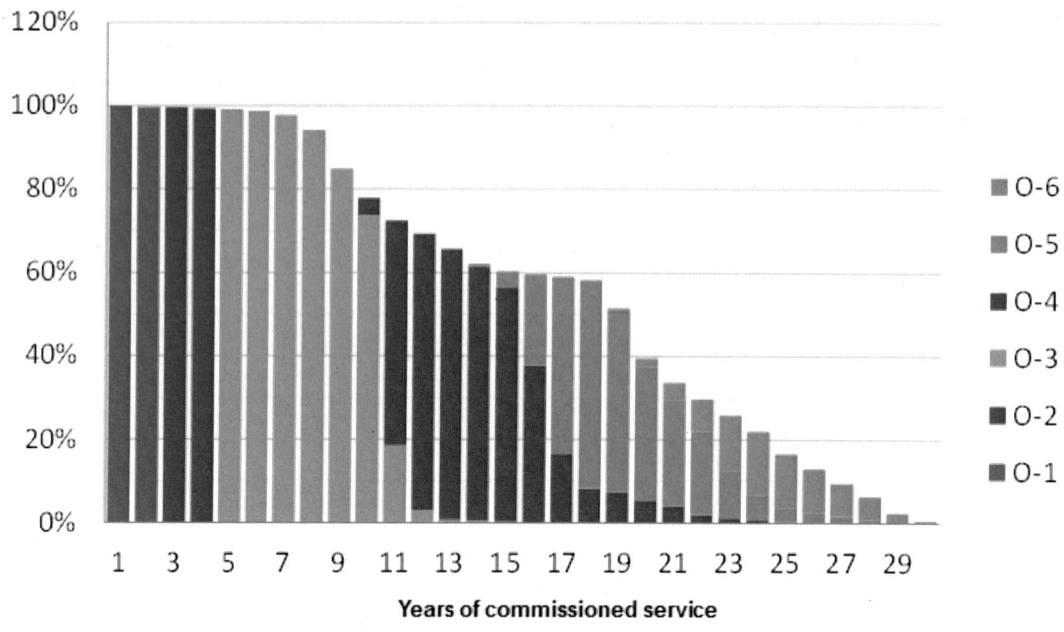

SOURCE: Air Force officer personnel records for 2001-2011.

---

[100] RAND gleaned the year-to-year transitions for individual officers during 2000-2011 from personnel records. As calculated for this analysis, each commissioned year of service's continuation rate is the fraction of the number of AC fighter pilots at the beginning of the year still in the AC at year's end, regardless of whether they were categorized as fighter pilots at that point.

Assessing Leadership Development Capacity

We used optimization and simulation models to tell how many officers the Air Force could develop using the numbers of jobs in Table 6.4 and the nominal career paths in Figure 6.2, guided by the sustainable experience/grade profile in Figure 6.3. The more officers that the AC can channel through the paths toward the left in Figure 6.2, the greater its *capacity* for developing fighter pilots with general officer or O-6 potential. The optimization model mathematically maximizes flows through those paths, also recommending average sojourns (cumulative times) in the corresponding job categories and grades, the sojourns constrained between specified upper and lower limits. We used a version that addresses only the field grades O-4, O-5, and O-6. It reflects only the combinations of jobs held within grade levels, not their sequences therein, insisting only that any jobs at O-4/5 precede any jobs at O-6. Its results represent upper limits on the numbers of officers who could be channeled through the various career paths.

We used a simulation approach to officer development to study what happens in a less carefully managed environment than the one assumed by the optimization model. We used a Monte Carlo simulation model known as the Military Career Model,[101] which represents the progression of simulated individuals through jobs and grades in a way similar to how the Air Force manages actual assignment and promotion processes. We configured it to model grades O-1 through O-6, with the jobs regarded as mandatory in any career path giving preference to candidates with prior experience in other mandatory and preferred jobs, with the preferred experience getting lower priority. Among other things, it simulates assignment cycles multiple times per year; assignments with minimum, maximum, and preferred durations; individuals with different retention likelihoods; thus reflecting considerably more detail and business rules than the optimization model. We configured the model in two ways for this analysis.

One way placed modeled officers in groups corresponding to the career paths in Figure 6.2 and managed their assignments and retention differentially. The other configuration did not differentiate, instead simply assigning the best-qualified and available officers at each cycle. Operated in the career-path configuration, the Military Career Model could produce very nearly as many officers with the mandatory experiences as the optimization model, establishing consistency between the models' notably different analytic approaches. We present the results from the non-career-path configuration because they more nearly reflect actual Air Force practice; they can be regarded as lower limits on the numbers of officers who could be channeled through the various career paths. The more the Air Force's assignment process took account of fighter pilots' future leadership potential, the closer it could come to realizing the optimization model's calculated production capacities.

---

[101] An early version of this model is documented in Schirmer (2009).

Figure 6.4 depicts one result from the optimization model, a maximization of the joint warfighter and organize/train/equip paths' total inputs and outputs (6.8 + 7.4 = 14.2 per year, as denoted at the bottom of the figure) that is necessary to, in the long term, provide the number of people required in each position for F-35 beddown Alternative 2A, denoted in the column on the right. For these calculations, we limited the sojourns for most jobs to no more than three or four years, and we required at least 1.5 or two years for the mandatory jobs. The jobs included in each career path show positive average sojourns—e.g., 1.5 years at Headquarters Air Force for O-4/5s on paths that would prepare them as candidates for senior leadership roles as joint warfighters, and two years for those grooming for organize/train/equip leadership at the general officer level. It is important to understand that the sojourn numbers represent averages: Some officers on a path could spend more than the average time in a job and others less—e.g., the average would be 1.5 years if all spent 1.5 years in a job category, or if half spent one year and the other half two years. For smaller numbers like the joint warfare path's average 0.9 years in wing/group/ Operations Support Squadron jobs at O-4/5, think of 90 percent of the path's officers spending one year in such jobs and 10 percent lacking that experience, or of 45 percent spending two years and 55 percent lacking the experience, and so on.

The Military Career Model turns out much more extensive assessments than the optimization, including a database detailing history over multiple simulated decades. For example, one can see how often jobs were filled by simulated officers with different levels of qualification (e.g., how many had at least two of a job's three preferred prior experiences). We summarized these results in terms of the numbers of "graduating" colonels with the combinations of jobs deemed mandatory for the various career paths (even though we didn't guide the model to channel officers through the career paths).

Figure 6.5 displays results from the optimization and simulation models, reflecting the total numbers of fighter-pilot colonels graduating with the combinations of experience deemed mandatory for Figure 6.2's joint warfare and organize/train/equip paths. The tops of the bars reflect the optimization's results and the bottoms the simulation's results. The simulation's annual production of fighter pilots with the experiences marked as mandatory on the two paths varies from 6.2 to 7.7 per year, a difference of 1.5 or about 24 percent. The flow optimization produces from 10.6 to 14.2 pilots per year, a difference of 3.6 or about 34 percent across these beddown alternatives. Given that five to seven fighter pilots per year have been promoted to general officer during recent years, these results suggest that the USAF will be somewhat constrained with respect to fighter pilot leadership development. To allow for a larger pool of candidates with the preferred characteristics, the USAF needs to be deliberate with its leadership development during the change from legacy fighter/attack aircraft to the F-35.

Figure 6.4. One Maximization of Annual Flows Through Two Development Paths

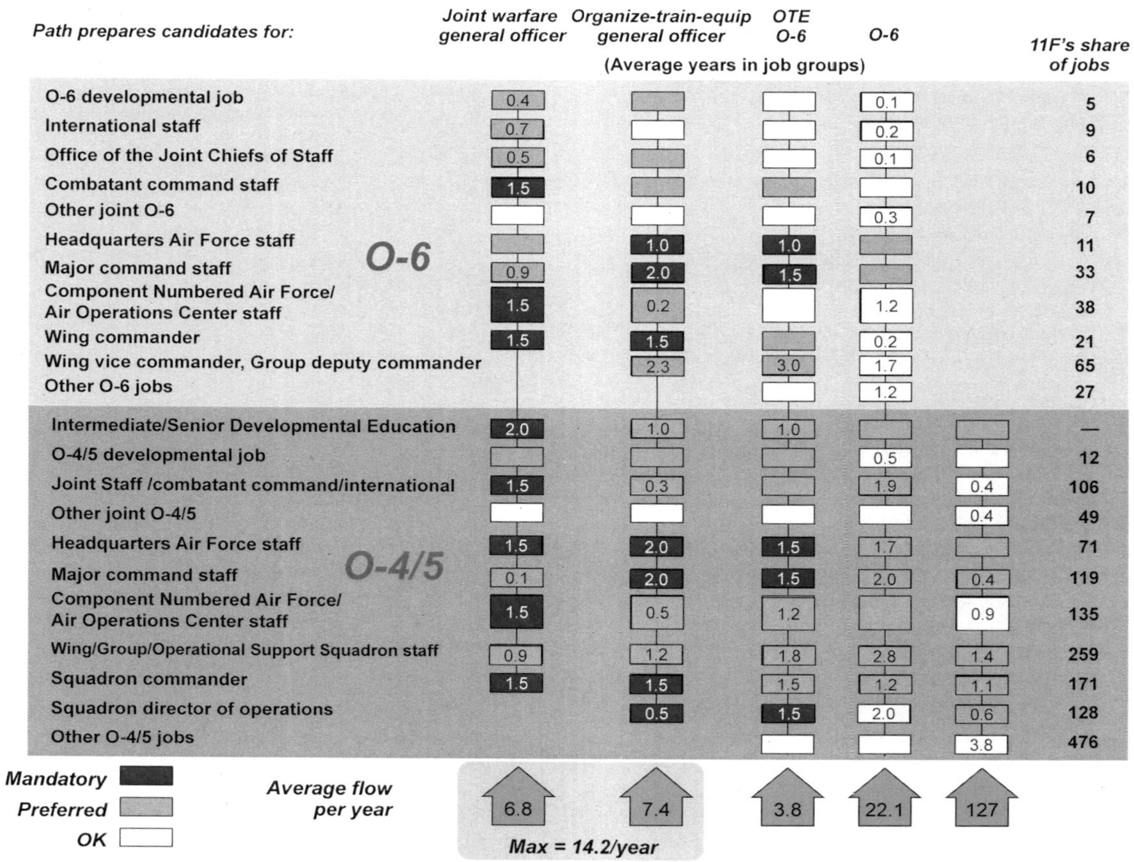

Figure 6.5. F-35 Beddown Alternatives Affect Leader Development by Less Than Assignment Policy

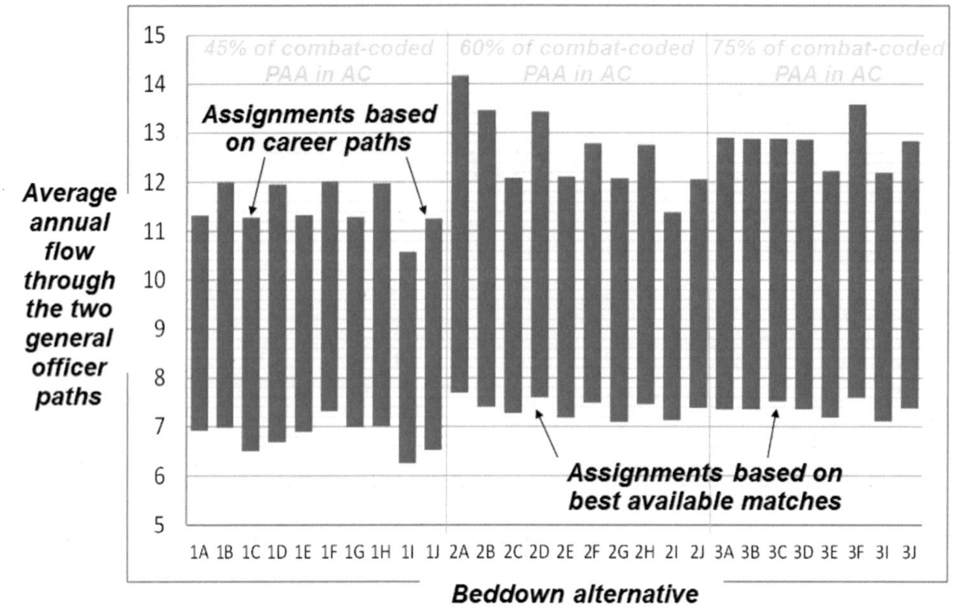

Note that the spans between the bottoms and tops of these bars range between 60 and 84 percent. That is, using planned career paths to guide assignments (i.e., building officers' combinations of experiences consistent with their perceived potential for future leadership roles) would have more effect than the F-35 beddown alternatives on the AC's production of candidates for leadership for these roles at the general officer level.[102]

All in all, we concluded that the F-35 beddown alternatives would have a slight effect on the AC's capacity for producing future senior leaders with targeted combinations of experience. However, these results suggest that the USAF will be somewhat constrained with respect to fighter pilot leadership development, aside from the impacts of squadron size. To allow for a larger pool of candidates with the preferred characteristics, the USAF needs to be deliberate with its leadership development during the change from legacy fighter/attack aircraft to the F-35, but none of the beddown alternatives with at least 60 percent of the combat-coded PAA in the AC would jeopardize its ability to produce at least as many well-qualified candidates as have actually been promoted to general officer during recent years.

[102] It is worth observing that other development/assignment policies can also exert greater effects than the F-35 beddown alternatives on the production of well-qualified candidates for senior leadership. For example, we found that allowing shorter sojourns in jobs that are in short supply (e.g., wing command positions) can increase capacity by more than the F-35 beddown alternatives.

7. Conclusions

This analysis has assessed whether O&S savings could be achieved by (1) reconfiguring the planned 960 USAF combat-coded F-35 PAA into larger squadrons, (2) adjusting the mix of PAA across the AC and RC, and (3) adjusting the percentage of the AC PAA assigned to CONUS home-station locations. Specifically, the research addressed how a change along these three dimensions would affect the Air Force's:

- Ability to support both surge and steady-state contingency operations
- Ability to absorb the necessary number of F-35 pilots
- Requirements for maintenance manpower and support equipment
- Requirements for new infrastructure across the set of existing F-16 and A-10 bases
- Ability to develop future senior leaders out of the pool of fighter pilots.

Within this analysis, we considered a set of 28 F-35 beddown alternatives. These alternatives vary across three dimensions. First, we considered three values for the percentage of total combat-coded F-35 PAA in the AC: 45 percent, 60 percent, and 75 percent. Second, we considered three values for the number of PAA per squadron in the AC: 24 PAA, 30 PAA, and 36 PAA. Within any beddown alternative, AFRC squadrons would always be assumed to have the same number of PAA per squadron as AC squadrons; however, only two values were considered for the number of PAA per ANG squadron: 18 PAA and 24 PAA. Third, we considered two values for the percentage of AC PAA that would be based at CONUS home-station locations: 50 percent and 67 percent.

Key Findings

Deployment Requirements

We found that all 28 beddown alternatives satisfy surge requirements. We examined the alternatives' ability to satisfy peak surge demand. All beddown alternatives were found to have sufficient squadrons to satisfy surge squadron requirements.

Further, most alternatives satisfy rotational requirements within specified deploy-to-dwell ratios. Squadron requirements for post-surge and non-surge periods were found to be sensitive to both AC/RC mix and to squadron size. We found that 18 out of 28 total beddown alternatives had sufficient squadrons to satisfy rotational requirements within the specified deploy-to-dwell ratios, and two additional beddowns could satisfy these requirements within the post-surge deploy-to-dwell ratios. Thus, most alternatives can satisfy rotational demands.

Pilot Absorption

Achieving feasible absorption conditions will require both a change in the burden historically borne by RC units and additional resources allowing AC units to overfly RAP minimums. Only one of the excursions analyzed (the first) produced pilot inventories that approached the required levels, and all of them tended to impose a disproportionate share of the absorption burden on the ANG and AFRC units. Indeed, the first absorption excursion tested required ANG unit UTE rates of two to three sorties per PAA per month higher than the AC UTE for the same beddown option, and ANG unit experience levels dropped below 60 percent for several beddown options in that excursion.

We found that squadron size and AC/RC mix affected experience levels in RC units; i.e., RC experience level increases with squadron size and with the percentage of aircraft in the AC. The RC UTE requirement to meet pilot absorption decreases as squadron size increases; this requirement was not significantly affected by the AC/RC mix. The AC UTE requirement decreases as the percentage of total aircraft in AC increases; this requirement was not significantly affected by squadron size.

These squadron size effects could have a significant impact on pilot absorption flying costs. As the fraction of combat-coded PAA in the AC is held constant, increasing the squadron size can significantly reduce the annual pilot absorption flying cost. Further, as the squadron size is held constant, increasing the fraction of combat-coded PAA in the AC also generates cost reductions. When compared to the ACC/CC-approved beddown's $4.4 billion in annual pilot absorption flying costs, there are alternative beddowns that satisfy all deployment requirements and reduce this cost by 10 percent or more.

Logistics Resources

Increasing squadron size reduces maintenance manpower requirements. For combat-coded aircraft, the required maintenance manpower per PAA decreases as the number of PAA per squadron increases. We estimated that a squadron of 36 PAA could be supported by 26 percent fewer maintenance positions per PAA than a squadron of 18 PAA. Further, assigning multiple squadrons to a single wing can generate additional savings beyond those generated by the squadron size effect. Our analysis suggests that a wing of three squadrons of 36 PAA each requires 6 percent fewer maintenance positions per PAA than a single squadron of 36 PAA.

Squadron size has a significant impact on total support equipment procurement costs. Increasing the squadron size and increasing the fraction of combat-coded PAA in the AC can significantly reduce the overall support equipment procurement cost.

Infrastructure

Utilizing current F-16 and A-10 bases, little additional capacity would be required under the 28 beddown alternatives. Some of the resource categories proved to be in sufficient supply for all bases under all beddown alternatives. In particular, for runway and ramp, no new capacity would be needed—all F-35 requirements can be satisfied with existing facilities. The other resource categories did require some additional capacity; however, in most cases these requirements are relatively small.[103]

The beddown alternatives also exhibited some cost reductions associated with consolidation to fewer bases. Larger squadron sizes reduce annualized facilities costs, while increasing the percentage of aircraft in the AC also reduces annualized facility costs.

Leader Development

Assignment policy had more effect on leader development than squadron size or AC/RC mix. All in all, we concluded that the F-35 beddown alternatives would have a slight effect on the AC's capacity for producing future senior leaders with targeted combinations of experience. However, these results suggest that the USAF will be somewhat constrained with respect to fighter pilot leadership development, aside from the impacts of squadron size. To allow for a larger pool of candidates with the preferred characteristics, the USAF needs to be deliberate with its leadership development during the change from legacy fighter/attack aircraft to the F-35, but none of the beddown alternatives with at least 60 percent of the combat-coded PAA in the AC would jeopardize its ability to produce at least as many well-qualified candidates as have actually been promoted to general officer during recent years.

Potential for Cost Reductions

Our primary finding is that increasing the F-35 squadron size from the levels utilized in the ACC/CC-approved beddown (24 PAA per AC and AFRC squadron, 18 PAA per ANG squadron) can satisfy both surge and steady-steady deployment requirements and can generate significant savings in the following areas:

---

[103] Note that this is based on analysis of raw square footage data from one OSD database and does not address the condition or adequacy of current facilities and infrastructure.

- Annual pilot absorption flying costs (more than $400 million)
- Annual maintenance manpower costs (more than $180 million)
- One-time support equipment requirements (more than $200 million)
- Annualized facilities costs (more than 10 percent).

The lower bounds on these estimates could be achieved, and all deployment requirements satisfied, if the USAF were to implement a posture that utilizes 30 PAA per AC and AFRC squadron and 24 PAA per ANG squadron (beddown alternatives 2G and 2H). The savings would increase if the USAF were to select a posture with 36 PAA in AC and AFRC squadrons and 24 PAA in ANG squadrons (Alternatives 2I and 2J), but this posture would assume increased risk; it has sufficient squadrons to satisfy surge wartime requirements, but it cannot satisfy steady-state requirements within the desired deploy-to-dwell ratios.

Further savings are possible in all categories except maintenance manpower, if the percentage of PAA in the AC were increased from the 60 percent level assumed in the ACC/CC-approved beddown. The percentage of AC PAA assigned to CONUS locations had little impact on these savings.

The Way Forward

These findings can be used to inform many issues that are within the purview of other USAF analyses and decision processes, including the Total Force Integration Roundtable's discussion of Associate Unit Force Presentation, the AF/A8X Multi-Role Fighter Phase II Force Composition Analysis, and the Strategic Basing Process performed by the SAF/IE and AF/A8. In particular, these findings can help determine how F-35 associate units should be composed and resourced in order to meet the requirements of increased pilot absorption and (potentially) increased deployment capability.

Bibliography

ACC — *See* Air Combat Command.

Air Combat Command, *Aircraft Maintenance Command Guide for ACC*, Langley Air Force Base, Virginia, May 2011.

AFM — *See* U.S. Air Force, Air Force Manual.

AFRCH — *See* U.S. Air Force, Air Force Reserve Command Handbook.

ANGH — *See* U.S. National Guard Bureau.

Anderegg, C.R., Sierra Hotel: *Flying Air Force Fighters in the Decade after Vietnam*, Washington, D.C.: U.S. Air Force History and Museums Program, 2001.

DoD — *See* U.S. Department of Defense.

Drew, John G., Kristin F. Lynch, James M. Masters, Robert S. Tripp, and Charles Robert Roll, Jr., *Options for Meeting the Maintenance Demands of Active Associate Flying Units*, Santa Monica, Calif.: RAND Corporation, MG-611-AF, 2008. As of October 24, 2012:
http://www.rand.org/pubs/monographs/MG611.html

Fisher, Rolland R., W. W. Drake, J. J. Delfausse, Andrew J. Clark, and Al Buchanan, *The Logistics Composite Model: An Overall View*, Santa Monica, Calif.: RAND Corporation, RM-5544-PR, 1968. As of October 24, 2012:
http://www.rand.org/pubs/research_memoranda/RM5544.html

Office of the Deputy Under Secretary of Defense, Installations and Environment, *Facilities Program Requirements Suite*, 2010. As of October 24, 2012:
http://www.acq.osd.mil/ie/fprs.shtml

Office of the Deputy Under Secretary of Defense, Installations an Environment, *Facilities Sustainment/Modernization/Operation Models*, 2011. As of October 24, 2012:
http://www.acq.osd.mil/ie/fim/programanalysis_budget/tool_metrics/FPRS/FSMDescription.shtml

McGarvey, Ronald G., James H. Bigelow, Gary James Briggs, Peter Buryk, Raymond E. Conley, John G. Drew, Perry Shameem Firoz, Julie Kim, Lance Menthe, S. Craig Moore, William W. Taylor, William A. Williams, *Assessment of Beddown*

Alternatives for the F-35: Executive Summary, Santa Monica, Calif.: RAND Corporation, RR-124/1-AF, 2012.

Moore, S. Craig, Brent Thomas, Raymond E. Conley, *Targeting the Occupational Skill Pairings Needed in New Air Force Colonels*, Santa Monica, Calif.: RAND Corporation, TR-759-AF, 2010. As of October 24, 2012: http://www.rand.org/pubs/technical_reports/TR759.html

Robbert, Albert A., Steve Drezner, John E. Boon Jr., Lawrence M. Hanser, S. Craig Moore, Lynn M. Scott, and Herbert J. Shukiar, *Integrated Planning for the Air Force Senior Leader Workforce: Background and Methods*, Santa Monica, Calif.: RAND Corporation, TR-175-AF, 2004. As of October 24, 2012: http://www.rand.org/pubs/technical_reports/TR175.html

Schirmer, Peter, *Computer Simulation of General and Flag Officer Management: Model Description and Results*, Santa Monica, Calif: RAND Corporation, TR-702-OSD. 2009. As of October 24, 2012: http://www.rand.org/pubs/technical_reports/TR702.html

Taylor, William W., James H. Bigelow, and John A. Ausink, *Fighter Drawdown Dynamics: Effects on Aircrew Inventories*, Santa Monica, Calif: RAND Corporation, MG-855-AF, 2009. As of October 24, 2012: http://www.rand.org/pubs/monographs/MG855.html

Taylor, William W., James H. Bigelow, S. Craig Moore, Leslie Wickman, Brent Thomas, and Richard S. Marken, *Absorbing Air Force Fighter Pilots: Parameters, Problems, and Policy Options*, Santa Monica, Calif: RAND Corporation, MR-1550-AF, 2002. As of October 24, 2012: http://www.rand.org/pubs/monograph_reports/MR1550.html

Thie, Harry J., William W. Taylor, Claire M. Levy, Sheila Nataraj Kirby, and Clifford M. Graff, II, *Total Force Pilot Requirements and Management: An Executive Summary*, Santa Monica, Calif: RAND Corporation, MR-646-OSD, 1995. As of October 24, 2012: http://www.rand.org/pubs/monograph_reports/MR646.html

USAF — *See* U.S. Air Force.

United Facilities Design Guide (UFC) 4-740-02 26, *Fitness Centers*, September 2006.

U.S. Air Force, Air Force Manpower Standard 21B1/21E1, *Maintenance Operations Flight/Current Operations Flight (Maintenance)*, Washington, D.C., March 21, 1997a.

———, Air Force Manpower Standard 23F1, *Aerospace Ground Equipment (AGE)*, Washington, D.C., May 16, 1997b.

———, Air Force Manpower Standard 22C1AC, *Operations Squadron Support Section*, Washington, D.C., June 13, 2002.

———, Headquarters, Office of the Civil Engineer, Housing Division (AF/ILEH), *Unaccompanied Housing Design Guide*, AFD-070919-082, January 2006. As of August 30, 2012: http://www.afcec.af.mil/shared/media/document/AFD-070919-082.pdf

———, Air Force Manpower Standard 10X0, *Group Commander's Support Staff*, Washington, D.C., October 16, 2008.

———, Air Force Instruction 11-401, *Aviation Resource Management Policies*, Washington, D.C., December 10, 2010.

———, Air Force Instruction 11-2F-16, Volume 1, *F-16 Pilot Training*, Washington, D.C., August 11, 2011a.

———, Air Force Instruction 38-201, *Management of Manpower Requirements and Authorizations*, Washington, D.C., September 26, 2011b.

———, Air Force Instruction 65-503, *U.S. Air Force Cost and Planning Factors, BY 2013 Logistics Cost Factors*, Washington, D.C., 2012a.

———, Air Force Manual (AFM) 32-1084, *Civil Engineering: Facility Requirements*, April 20, 2012b.

U.S. Department of Defense, *Selected Acquisition Report (SAR): F-35 Joint Strike Fighter*, December 31, 2011.

U.S. National Guard Bureau, *Air National Guard Handbook* (ANGH) 32-1084, Facility Space Standards, April 2, 2012.